D1827299

Mathematical Engineering

Series editors

Jörg Schröder, Essen, Germany
Bernhard Weigand, Stuttgart, Germany

Today, the development of high-tech systems is unthinkable without mathematical modeling and analysis of system behavior. As such, many fields in the modern engineering sciences (e.g. control engineering, communications engineering, mechanical engineering, and robotics) call for sophisticated mathematical methods in order to solve the tasks at hand.

The series Mathematical Engineering presents new or heretofore little-known methods to support engineers in finding suitable answers to their questions, presenting those methods in such manner as to make them ideally comprehensible and applicable in practice.

Therefore, the primary focus is—without neglecting mathematical accuracy—on comprehensibility and real-world applicability.

To submit a proposal or request further information, please use the PDF Proposal Form or contact directly: *Dr. Jan-Philip Schmidt, Publishing Editor (jan-philip. schmidt@springer.com).*

More information about this series at http://www.springer.com/series/8445

Stefan Diebels · Sergej Rjasanow
Editors

Multi-scale Simulation of Composite Materials

Results from the Project MuSiKo

 Springer

Editors
Stefan Diebels
Technische Mechanik
Universität des Saarlandes
Saarbrücken, Germany

Sergej Rjasanow
Universität des Saarlandes
Saarbrücken, Germany

ISSN 2192-4732   ISSN 2192-4740   (electronic)
Mathematical Engineering
ISBN 978-3-662-57956-5   ISBN 978-3-662-57957-2   (eBook)
https://doi.org/10.1007/978-3-662-57957-2

Library of Congress Control Number: 2018966395

© Springer-Verlag GmbH Germany, part of Springer Nature 2019
This work is subject to copyright. All rights are reserved by the Publisher, whether the whole or part of the material is concerned, specifically the rights of translation, reprinting, reuse of illustrations, recitation, broadcasting, reproduction on microfilms or in any other physical way, and transmission or information storage and retrieval, electronic adaptation, computer software, or by similar or dissimilar methodology now known or hereafter developed.
The use of general descriptive names, registered names, trademarks, service marks, etc. in this publication does not imply, even in the absence of a specific statement, that such names are exempt from the relevant protective laws and regulations and therefore free for general use.
The publisher, the authors and the editors are safe to assume that the advice and information in this book are believed to be true and accurate at the date of publication. Neither the publisher nor the authors or the editors give a warranty, express or implied, with respect to the material contained herein or for any errors or omissions that may have been made. The publisher remains neutral with regard to jurisdictional claims in published maps and institutional affiliations.

This Springer imprint is published by the registered company Springer-Verlag GmbH, DE part of Springer Nature
The registered company address is: Heidelberger Platz 3, 14197 Berlin, Germany

Preface

Due to their high stiffness and strength and their good processing properties, short fibre reinforced thermoplastics are well-established construction materials.

Up to now, simulation of engineering parts consisting of short fibre reinforced thermoplastics has often been based on macroscopic phenomenological models, but deformations, damage and failure of composite materials strongly depend on their microstructure. The typical modes of failure of short fibre thermoplastics enriched with glass fibres are matrix failure, rupture of fibres and delamination, and pure macroscopic consideration is not sufficient to predict those effects. The typical predictive phenomenological models are complex and only available for very special failures. A quantitative prediction on how failure will change depending on the content and orientation of the fibres is generally not possible, and the direct involvement of the above-mentioned effects in a numerical simulation requires multi-scale modelling. On the one hand, this makes it possible to take into account the properties of the matrix material and the fibre material, the microstructure of the composite in terms of fibre content, fibre orientation and shape as well as the properties of the interface between fibres and matrix. On the other hand, the multi-scale approach links these local properties to the global behaviour and forms the basis for the dimensioning and design of engineering components. Furthermore, multi-scale numerical simulations are required to allow for an efficient solution of the models when investigating three-dimensional problems of dimensioning engineering parts.

Bringing together mathematical modelling, materials mechanics, numerical methods and experimental engineering, this book provides a unique overview of multi-scale modelling approaches, multi-scale simulations and experimental investigations of short fibre reinforced thermoplastics. The first chapters focus on two principal subjects: the mathematical and mechanical models governing composite properties and damage description. The subsequent chapters present numerical algorithms based on the Finite Element Method and the Boundary Element Method, both of which make explicit use of the composites microstructure. Further, the results of the numerical simulations are shown and compared to experimental results.

Lastly, the book investigates deformation, fatigue and failure of composite materials experimentally, explaining the applied methods and presenting the results for different volume fractions of fibres.

This book is a valuable resource for applied mathematicians, theoretical and experimental mechanical engineers as well as engineers in industry dealing with modelling and simulation of short fibre reinforced composites.

This research was supported by the Federal Ministry of Education and Research (BMBF) of Germany within the BMBF programme "Mathematics for Innovations in Industry and Services, 2013–2016".

Saarbrücken, Germany   Stefan Diebels
November 2018   Sergej Rjasanow

Contents

Chapter 1
Multi-Scale Methods in Simulation—A Path to a Better Understanding of the Behaviour of Structures

Michael Hack

1.1 State of the Art—What We Can Do Today

The success of using simulation methods is highly connected to the efficiency one can use them. Efficiency depends on many factors like process factors (Fig. 1.1):

- Easiness of setting up the simulation,
- Seamless process integration: applying the correct data at the correct places,
- Effectiveness of interpreting the results of the simulation,
- Integration into optimisation runs.

For these processes and user interaction oriented tasks, the software companies tailor and provide their software platforms for the needs of their applications and users, see e.g. [1].

But there are also strong influences of the methodology on accuracy and efficiency of simulation tools:

- Get the correct material data,
- Adaptive calculation,
- Parallel computing.

To be able to improve the latter, a close collaboration between industry and research is necessary. For the industry it helps to get access the most efficient methods available, and for research it ensures that the newest developments find their way into real applications.

Here applied mathematics plays a crucial role to provide deep insights into the algorithms and applicability. So it is no surprise that the subsidiary of Siemens PLM for durability in Kaiserslautern was originally a spin-off of the Mathematics Department at the TU Kaiserslautern.

M. Hack (✉)
Siemens PLM Software, Kaiserslautern, Germany
e-mail: michael.hack@siemens.com

© Springer-Verlag GmbH Germany, part of Springer Nature 2019
S. Diebels and S. Rjasanow (eds.), *Multi-scale Simulation of Composite Materials*,
Mathematical Engineering, https://doi.org/10.1007/978-3-662-57957-2_1

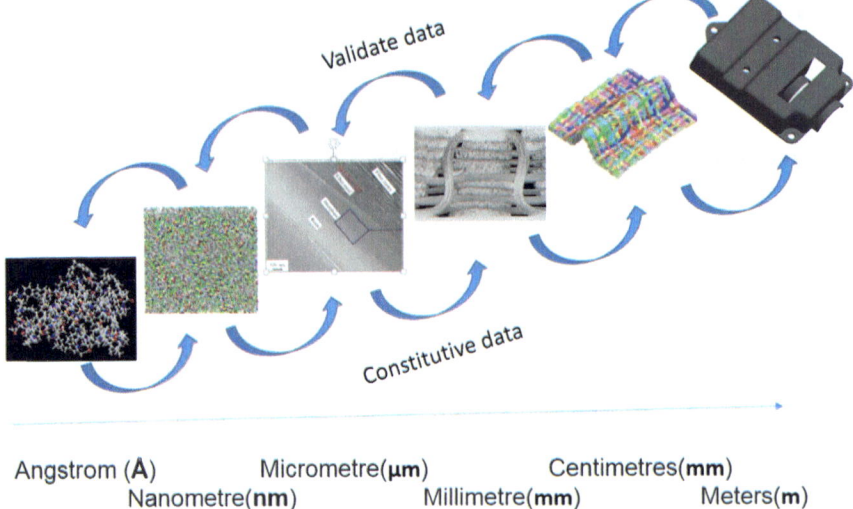

Angstrom (**Å**)   Micrometre(**μm**)   Centimetres(**mm**)
        Nanometre(**nm**)   Millimetre(**mm**)   Meters(**m**)

Fig. 1.1 In multi-scale approaches the smaller scales provide the information that the next level needs to run its analysis. Anything not coming from the scale below needs to be achieved by tests on the upper scale (Figures from [2–6])

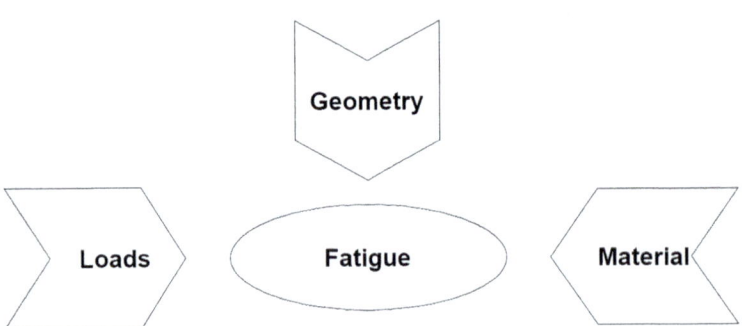

Fig. 1.2 Traditional view on the influences on fatigue life

Over the last decades new and highly efficient methodologies could find their way into software. More than twenty years ago fatigue simulation with realistic loads and complex structures could take weeks to run—no real time gain against testing. As usage in the daily design process was not possible, the loads typically were simplified, and very special load conditions were analysed. Another approach was to restrict the analysis to user defined points. As these approaches could not esti-mate the error introduced by the simplifications, the user had still to live with large safety factors and over-designs. In a joint project new filter tools had been developed in a Ph.D.-thesis [7] that provided good error estimates and efficient implementa-tion. Using this approach, real live problems could be solved within hours, see [8] (Fig. 1.2).

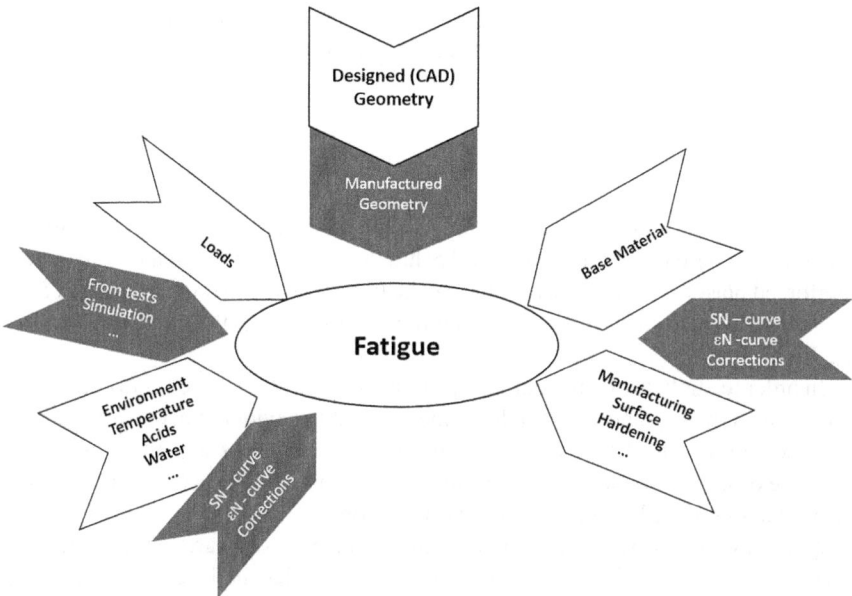

Fig. 1.3 In reality much more factors especially from the manufacturing process are acting. Traditionally they are captured by correction factors

1.2   Requirements from the Applications—New Materials Need New Methodologies

In metals the fatigue problems typically initiate on imperfections in the metal or on grain boundaries. These micro-structures are influenced by the manufacturing process but the local distribution is normally unknown. On the other hand, two centuries of fatigue testing on metal structures have given enough background to replace a real modelling of the micro-structure by test based material data. This was, and still is, possible as the material behaviour is quite homogeneous, and local behaviour can be handled by empirical correction factors, see Fig. 1.3. Once we face composite structures, we get a much more complex situation. While in metals the damage mechanism is more or less unique namely the initiation of micro-cracks, in composites there are multiple mechanisms as multiple materials are involved. Also in metals the behaviour is typically isotropic and the local anisotropic behaviour can again be handled by multi-axial criteria [8]. In composite structures the damage mechanisms are more complex and may influence each other.

So trying to follow the empirical based approach, like for metals to get material data, would at least lead to a tremendous set of test setups and also need so many simplifications due to the missing reliability in material data large safety factors would be needed.

In the next sub-sections, we depict the challenges for short fibre reinforced plastic materials analysed in the studies of this book project.

1.2.1   Orientation of Fibres

For composite structures, it is necessary to take the local microscopic structure into account, as it defines the basic structural behaviour. For injection moulded short fibre reinforced plastics the local distribution of the fibre orientation directly influences the (anisotropic) local stiffness of the structure and also defines the damage behaviour, see Fig. 1.1.

In order to study the behaviour of injection moulded short fibre reinforced plastic components the evaluation of the local mechanical behaviour is necessary. In a first step the simulation of the manufacturing process itself has to be conducted. An outcome of such a simulation is the local distribution of the fibres (to be more precise the probability distribution of the fibre orientations).

Nowadays the usage of injection moulding simulation software is a standard in the development process. Such simulation software provides the local fibre orientation distribution as an output. The fibre orientation subsequently has to be transferred from the model for the injection moulding simulation into the model for the structural simulation.

1.2.2   One Approach—The Master SN Curve Approach

The local distribution of fibres lead to different fatigue behaviour at each point of the structure. The behaviour also changes with the direction the load is applied. Therefore it is necessary that for a given material the fatigue behaviour is known for any fibre distribution and any direction with respect to applied loads.

One idea to estimate the local material data is to combine test based approaches with simulation approaches on the micro-scale. In a joint research project, the KU Leuven and Siemens developed a hybrid master SN-curve approach [6, 9, 10]. The basic idea is to separate the influences of the orientation from the basic fatigue behaviour of the material (incl. considering temperature, wetness, etc.).

It was found that taking into account the effect of fibre matrix debonding and fibre cracking on micro-level enables to split the effect of the fibre orientation from the effects of the base materials and the environment, see [9, 10].

The basic idea behind the approach is to calculate the damage on microscopic level. The first cycle of loading is modelled. The onset of debonding, progression of fibre matrix debonding, and the subsequent loss of stiffness is based on the concept of *equivalent bonded inclusions*. A thorough mathematical treatment of this concept has been presented in [11].

In order to calculate a point on a SN-curve for an arbitrary but given orientation, one starts from the point of same (macroscopic) damage on a given master SN-curve

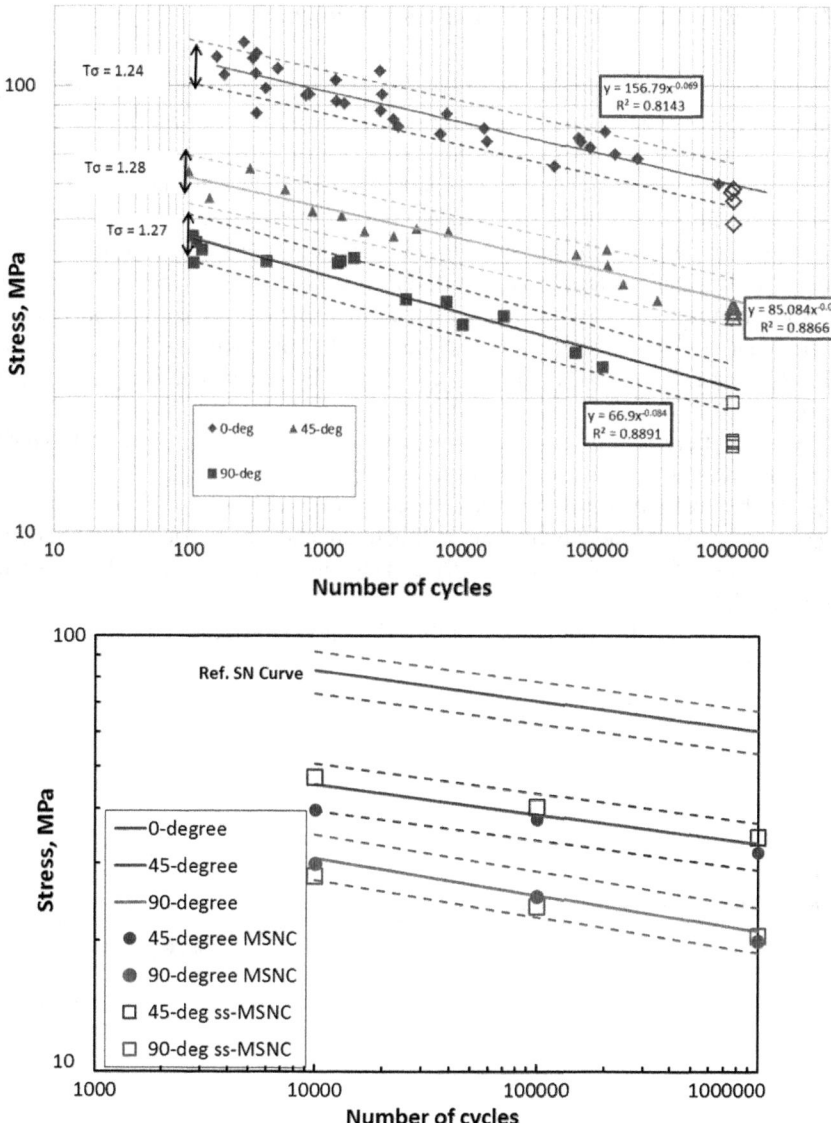

Fig. 1.4 Results of the master SN-curve approach (lower picture) compared to test results (upper picture): The predicted points lie within the scatter band of the tests [6]

and calculates the progressive damage on the microscopic scale for the orientation of the SN-curve. In order to get the point on the new SN-curve, the load that is needed to reach the same microscopic damage for the new orientation is evaluated by an iterative process. In Fig. 1.4 the results for a 50% Glass fibre reinforced PBT are shown.

1.2.3 Local and Global Stiffness Reduction

As opposed to metal structures, in composite structures a change in the local and
global stiffness before failure of the complete structure is observed. It can be seen
in the matrix material as well as in the composite structure. Detailed analysis on
different specimen had shown that this stiffness reduction over the lifetime is (at
least statistically) independent of the local fibre orientation [9].

These local stiffness changes lead to a redistribution of stresses. The influence
of these re-distributions lead to large differences between the component behaviour
and specimen behaviour. The slopes of specimen SN-curves are typically much
smaller than those of the component SN-curves. Without taking stiffness reduction
into account a correct simulation of the component behaviour was not possible, see
[6, 9].

For short fibre reinforced composites an exponential decay down to 90–85%
during the lifetime gives a good estimate, see Fig. 1.5.

To be able to apply a stiffness reduction algorithm with complex load scenarios,
it is necessary to use mathematical modelling to understand the background fatigue
simulation tools for metals and correctly enhance them for the needs of composite
fatigue modelling.

In the case of variable amplitude, traditional fatigue approaches for metallic material use SN-curves, linear Miner-Palmgren [12] damage accumulation and cycle
count (rainflow counted cycles [13]) based damage evaluations.

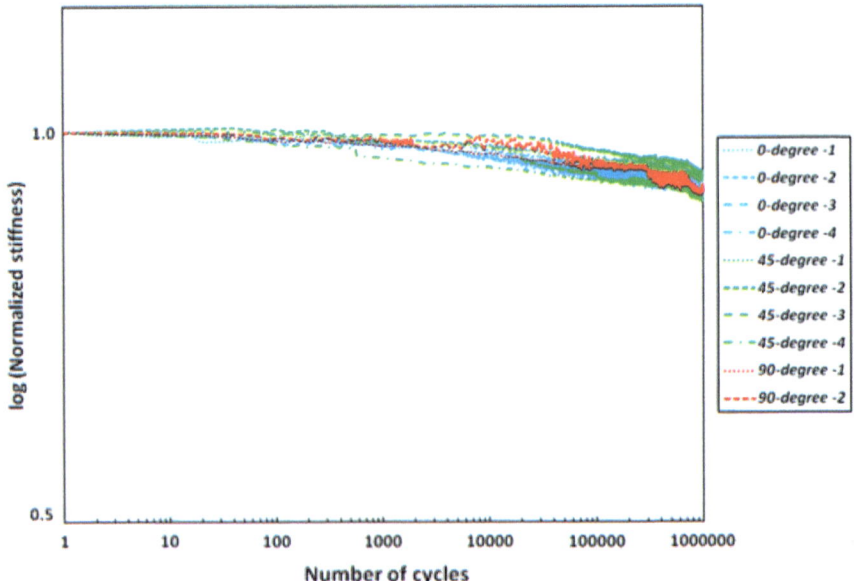

Fig. 1.5 Decay of stiffness for short fibre reinforced specimen under cyclic loading [6]

In 1945, Miner developed a linear damage accumulation method, based on the work of Palmgren and added the contribution of various stress amplitude loading to the damage. However, as for SN-curves, the loading history of the material is not accounted for. In rainflow counting methods the damage level depend on full closing hysteresis loop of load cycles. In the case of composite materials, the fatigue behaviour is changing over time due to changes in the matrix damage state. When applying variable amplitude loading, the largest load cycles that contribute to the larger amount of damage commonly take a very long time to complete, due to the many nested cycles. In this case, the approach to only consider cycles when they are completed can no longer be justified.

In the 1990 Brokate and Krejci [14] applied the mathematical tools of hysteresis operators to fatigue theory [15] analysing the linear damage accumulation and analogies between damage accumulation and energy dissipation.

Based on this work, it is possible to extend the rainflow based methodology to non-linear damage accumulation in a both mathematical and methodological sense: the damage hysteresis operator approach [7, 16].

The idea is based on the hysteresis operators for kinematic hardening (i.e. to calculate elastic-plastic stress-strain behaviour from pseudo elastic stress histories) and how dissipated energy is calculated in these models. The new idea is to replace the constitutive laws of elasto-plastic stress-strain behaviour with constitutive laws for a stress-damage potential behaviour [14, 15].

The hysteresis operator approach is able to calculate damage at any time increment instead of closed cycle increment. The extensions explained in [7, 16] allow the damage status and the damage behaviour to be updated depending on internal (i.e. pre-damage) and any external factors (i.e. temperature, humidity). Therefore this approach is also suited to follow the progressive damage curves and also including the damage history of the material.

Combining these methodology it is possible to simulate the fatigue life for injection moulded structures with limited effort in testing [6, 17].

1.3 Open Tools—Necessary for Including New Methodologies

We already learned in the sections before that for efficient and accurate simulation the methodology and process needs to be adapted to the

- material as is,
- manufacturing processes—i.e. material as manufactured,
- environmental factors and loads,
- pre-damages and actual local damages due to the load,
- and many more.

MaBIFF: **FKM-guideline:**

VL.Durability: endurance limit: **190 – 240 MPa**   VL.Durability: endurance limit: **140 MPa**

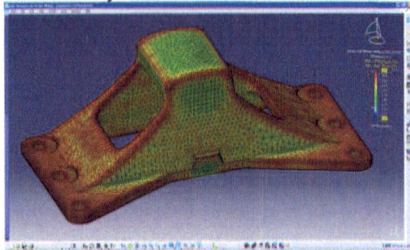

Fig. 1.6 Comparison of the allowable stress values (shift in SN-curves): In the left picture individual allowable stresses including the manufacturing simulation [19], on the right traditional approach using standards [18]

1.3.1 Manufacturing Influences

In Sect. 1.2 we have seen that classically, fatigue data have been taken from material tests for the material as available for the testing. In reality, the material as manufactured often has different properties. Indeed in many cases special treatment is added to manufacturing process to improve the properties of the material in places where the structure has to endure higher loads. The traditional approach to incorporate this influences in a fatigue simulation is to apply correction factors to modify the fatigue data (typically the SN-curve) [18].

From an application point of view this can get to a tedious process, as the influencing factors have to be extracted from the results of special manufacturing simulation tools. From those the correction factors have to be calculated (often done with spreadsheet calculation software) and applied to the fatigue simulation tools. The latter often uses different representations (FE-Meshes) of the structure and so even the simple application of the correction factors at the correct places (nodes/elements) is an error-prone process.

To establish more applicable processes to include manufacturing processes the BMBF-funded project MaBIFF [19] included this task for the manufacturing process of casting. In this project, first an investigation of the different properties of the casted material on a micro-scale (22 properties on the distribution, structure, and size of graphite, perlite, and ferrite) were analysed both from tests (micro-graphs) and simulation. In a second step, the micro-scale properties were correlated with fatigue data on the macro-scale.

For a real structure, first a simulation of the micro-scale material data was performed. A mapping used the correlated fatigue data directly in the fatigue solver. The fatigue solver for this was enhanced (opened) to be able to directly take the fatigue data directly from the process. The project showed for the analysed component that at the most loaded areas the manufacturing process lead to much better fatigue properties than for the rest of the structure and even better compared to standard fatigue data from literature, see Fig. 1.6.

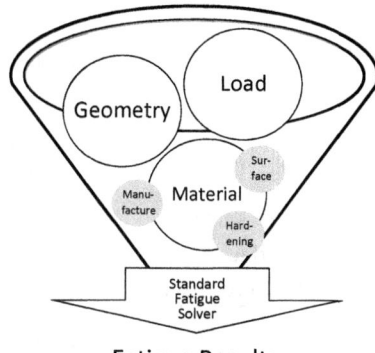

Fig. 1.7 Traditional approaches to incorporate influences into fatigue methodology: To be able to include it the influences where just added as correction factors to the material data, Hence the solver was a kind of bottle neck

1.3.2 New Methodologies

While for metallic structures, the methodology for fatigue is established and also the influence from manufacturing has widely been included in these methods by adapting the parameters, composite materials show more complex damaging behaviour. Often different damage mechanisms act at the same time: in material damage (matrix cracks, fibre cracking) but also interface damage (debonding) and interaction to other materials or plies (delamination). They may not only occur at the same time but also interact. The different mechanisms and the importance of them depend on many factors:

- Selection of combined materials,
- Volume fraction of the different materials,
- Geometry of the materials (Fibre length, aspect ratio, …),
- Topology of the materials,
- External factors like temperatures and humidity.

This means that a lot of different methodologies for different materials but maybe also for different levels of detail exist (Fig. 1.7).

Instead of implementing loads of different tools for each application and material the better way is to allow one solver tool to integrate different methodologies into one solver and only exchange parts of the implementation:

- how to filter the important areas in the structure and the important parts of the loads for efficiency,
- how to get from local finite element results to local input for damage and fatigue (forces, energies, stresses, strains),
- how to increase different damage modes,
- what is the influence on local stiffness,
- when local failure occurs (Fig. 1.8).

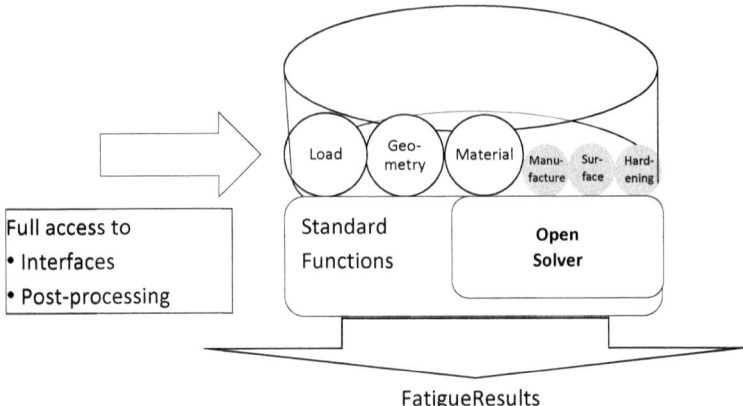

Fig. 1.8 In the open solver approach [1] the new influence factors are on the same level as traditional influence factors. They may also lead to changes in the solver, so we can go for any and the locally best methodology

This approach has already been successfully implemented for applications like

- Intra-laminar damage in continuous fibre structures, [20–22],
- Short-fibre reinforces plastics [6, 17],
- Adhesives.

1.4 On the Path with MuSiKo

In the project MuSiKo, as presented in this book, many of the efficiency aspects and the process aspects have been analysed in detail and with mathematical rigorousness for the application of damage in short fibre reinforced plastics. An important point in the project was also that it did not just focus on mathematical methods, but also, on the testing side, on how to get parameters to fill in the methods.

In all this means the study perfectly fits in the projects and developments mentioned in the sections above. Whereas in the examples, the influence on micro-scale is in most parts accounted for by a homogenisation to macro level on a testing based or at least hybrid approach, here a full scale of methods from empirical to full FE^2 is analysed.

Even though the main application case is static damage and not fatigue damage, results are important for the whole process.

Especially the filter methodology is of interest for many cases of fatigue and damage analysis, as they are not restricted to the application case of short fibre reinforced composites.

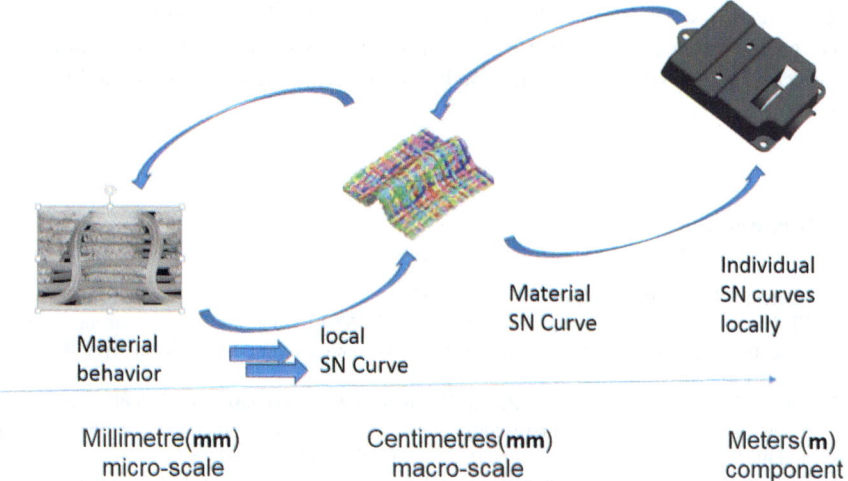

Fig. 1.9 In the past the step from marco-level to component level was established by the strain-life methods and the introduction of FE-methods. For composites we need the same step from micro-level to macro-level

1.5   Outlook—Virtual Testing and Material Design

We have seen that the applicability of simulation methods depends on the availability of the data that are needed to run them. Looking historically, there are several milestones in the fatigue of metals. Starting in the 19th century with component tests [23] that allowed to simulate the accumulation, the damage induced by realistic loads for a given structure, typically rail axles. An important step was to get from the component level to the local analysis. Nominal stress based and local strain-life approaches were developed. Here, material based fatigue data was needed. With the material based data and the correction mechanisms, it is possible to analyse arbitrary structures with these approaches with a limited set of test data. The scale here is what we call macro-scale today. When we look at the composite structures a similar step is needed. The material data needed at macro-level can hardly be tested. Again the goal is to start from basic data, the best that can be determined on the basic constituents of the material the matrix and the fibres. The goal is to be able to only use data on the material and the manufacturing process for estimating the behaviour of the material on macro level (Fig. 1.9).

We have seen there are different approaches to get the needed data. In Sect. 1.2, we describe a hybrid approach where data on micro level are combined with results of manufacturing simulation and test data on macro level. In [24], methods had been developed to bypass the tests on macro level and estimate those from data on the material level only.

The MuSiKo project, as described in this book, analyses several mathematical and numerical building blocks of methods that can be applied on micro-level directly, as well as how to derive the needed data from tests. Even though some of the methods may today still seem hard to use from an point of view of computation time, it is quite clear that several of the ideas will also progress the multi-scale methodology for damages in the future.

References

1. Siemens PLM Software: Simcenter 3D (2017)
2. El Said, B., Ivanov, D., Long, A.C., Hallett, S.R.: Multi-scale modelling of strongly hetero-geneous 3D composite structures using spatial Voronoi tessellation. J. Mech. Phys. Solids **88**, 50–71 (2016)
3. Wu, Q., Li, M., Gu, Y., Wang, S., Zhang, Z.: Imaging the interphase of carbon fiber composites using transmission electron microscopy: preparations by focused ion beam, ion beam etching, and ultramicrotomy. Chin. J. Aeronaut. **28**(5), 1529–1538 (2015)
4. Seveno, D., Van Duin A.: Curing of Epoxy Resins: A Nanoscale View. Copenhagen, Denmark (2015)
5. Hadden, C.M., Jensen, B.D., Bandyopadhyay, A., Odegard, G.M., Koo, A., Liang, R.: Molecular modeling of EPON-862/graphite composites: interfacial characteristics for multiple crosslink densities. Comput. Sci. Technol. **76**, 92–99 (2013)
6. Jain A.: Hybrid multi-scale modelling of damage and fatigue in short fiber reinforced composites. Ph.D. Thesis, Leuven (2015)
7. Hack, M.: Schädigungsbasierte Hysteresefilter. Ph.D. Thesis, TU, Kaiserslautern (1998)
8. Brune, M., Fiedler, B., Köttgen, V.B., Reißel, M.: FEM Based Durability Analysis of the Knuckle of the 5. Series BMW, Fatigue Design 98, Helsinki (1998)
9. Jain, A., Abdin, Y., Straesser, S., Van Paepegem, W., Verpoest, I., Lomov, S.V.: Master SN-curve approach—a hybrid multi-scale approach to fatigue simulation of short fiber composites. Compos. Part A Appl. Sci. Manuf. **91**, 510–518 (2016)
10. Jain, A., Abdin, Y., Van Paepegem, W., Verpoest, I., Lomov, S.V.: A feasibility study of the Master SN curve approach for short fiber reinforced composites. Int. J. Fatigue **91**(1), 264–274 (2016)
11. Jain, A., Abdin, Y., Van Paepegem, W., Verpoest, I., Lomov, S.V.: Effective anisotropic properties of inclusions with imperfect interface for Eshelby-based models. Compos. Struct. **131**, 692–706 (2015)
12. Miner, M.A.: Cumulative damage in fatigue. J. Appl. Math. **12**, A159–A164 (1945)
13. Matsuishi, M., Endo, T.: Fatigue of metals subjected to varying stresses. Jpn. Soc. Mech. Eng. Jukvoka (1968)
14. Brokate, M., Sprekels, J.: Hysteresis and Phase Transitions. Springer (1996)
15. Brokate, M., Dressler, K., Krejci, P.: Rainflow counting and energy dissipation for hysteresis models in elastoplasticity. Eur J. Mech A/Solids **4**, 705–737 (1996)
16. Nagode, M., Hack, M.: The damage operator approach: fatigue, creep and viscoplasticity modeling in thermo-mechanical fatigue, SAE int. J. Mater. Manuf. **4**, 632–637 (2011)
17. Hack, M., Korte, W., Sträßer, S., Teschner, M.: Fatigue simulation of a short fiber reinforced oil-filter under high temperature and pressure loads. Procedia Eng. **213**C, 207–214 (2018)
18. FKM: Analytical Strength Assessment, 6th ed. VDMA (2013)
19. Hack, M., Jung, D., Egner-Walter, A.: Optimized analysis process for the fatigue analysis of cast iron structures taking into account the local material structure. Mater. Test. **54**(7–8), 497–502 (2012)
20. Carrella-Payan, D., Magneville, B., Hack, M., Naito, T., Urushiyama, Y., Yamazaki, T., Van Paepegem, W.: Implementation of fatigue model for unidirectional laminate based on finite element analysis: theory and practice. Frattura ed Integrita Strutturale **38**, 184–190 (2016)

21. Hack, M., Carrella-Payan, D., Magneville, B., Naito, T., Urushiyama, Y., Yamazaki, T., Van Paepegem, W.: A progessive damage fatigue model for unidirectional laminated composites based on finite element analysis. Frattura ed Integrita Strutturale **44** (2018)
22. Carrella-Payan, D., Magneville, B., Naito, T., Yamazaki, T., Allaer, K., Van Paepegem, W., Matveeva, A., Hack, M.: Parameter Identification of delamination onset in mode I (UD and Angled-ply interface) for static and fatigue simulation of CFRP. Proc. Fatigue Des. (Senlis) **17** (2017)
23. Wöhler, A.: Über die Festigkeitsversuche mit Eisen und Stahl. Bauwesen **20**, 73–106 (1870)
24. Abdin Y.: Micro-mechanics based fatigue modelling of composites reinforced with straight and wavy short fibers. Ph.D. Thesis, Leuven (2015)

Chapter 2
Indicators for the Adaptive Choice of Multi-Scale Solvers Based on Configurational Mechanics

Ralf Müller, Charlotte Kuhn, Markus Klassen, Heiko Andrä and Sarah Staub

2.1 Introduction

Configurational mechanics deals with the treatment of driving forces on different
types of defects and inhomogeneities. Classically defects and inhomogeneities
can be characterised by their dimensionality. In a three dimensional continuum,
point defects, such as interstitial atoms and vacancies, are zero dimensional defects.
Dislocation lines or crack fronts represent one dimensional defects. Two dimensional
defects are interfaces such as phase boundaries. Continuously distributed material
properties in gradient materials can be considered as three dimensional inhomogeneities. The theory of configurational forces provides a unified approach to characterise all different types of defects and inhomogeneities. In a general approach,
configurational forces can be seen as driving forces on these different defects and
inhomogeneities. Configurational forces are of energetic character and are work
conjugated to the motion of defects and inhomogeneities. The general theory of configurational forces can be found in the books of [7, 12, 14, 15]. The derivations are
different, but they all agree on the final result.

R. Müller (✉) · C. Kuhn
University of Kaiserslautern, 67653 Kaiserslautern, Germany
e-mail: ram@rhrk.uni-kl.de

C. Kuhn
e-mail: chakuhn@rhrk.uni-kl.de

M. Klassen
RWTH Aachen University, 52074 Aachen, Germany
e-mail: klassen@lbb.rwth-aachen.de

H. Andrä · S. Staub
Fraunhofer Institute for Industrial Mathematics, 67663 Kaiserslautern, Germany
e-mail: heiko.andrae@itwm.fraunhofer.de

S. Staub
e-mail: sarah.staub@itwm.fraunhofer.de

© Springer-Verlag GmbH Germany, part of Springer Nature 2019
S. Diebels and S. Rjasanow (eds.), *Multi-scale Simulation of Composite Materials*,
Mathematical Engineering, https://doi.org/10.1007/978-3-662-57957-2_2

15

In numerical schemes configurational forces also appear if the classical field equations (equation of motion, kinematics, and constitutive equations) are not satisfied exactly, but only within the approximation of the numerical scheme, e.g. equilibrium only in a weak sense in finite element schemes. This allows for the use of configurational forces as a basis to improve the discretisation, for r-adaptive procedures see e.g. [1, 18, 19, 21, 27–29] or for h-adaptive schemes see [8, 20].

In the context of multi-scale approaches configurational forces were used to transfer information of the defect state on a micro-level to the macro-level. As shown in [11, 13, 25, 26] the inhomogeneity on the micro-level causes configurational forces that can be homogenised and influence e.g. the crack propagation on the macro-level. It must be mentioned that the homogenisation of configurational forces is not straightforward. The following section will be devoted to the discussion of the details of homogenisation of configurational forces and the proper associated scale transitions. To make the derivation compact and self contained, we will first briefly recall the theory of configurational forces in a small strain framework. An extension to finite strains can be found in [25, 26].

2.2 Basics of Configurational Mechanics

We restrict attention to static problems in the infinitesimal strain setting for an elastic material, where the field equations are given by:

$$\text{equilibrium: } \operatorname{div} \boldsymbol{\sigma} + \boldsymbol{f} = \mathbf{0},$$

$$\text{kinematics: } \boldsymbol{\varepsilon} = \operatorname{grad}_{\text{sym}} \boldsymbol{u} = \frac{1}{2}\left(\operatorname{grad} \boldsymbol{u} + (\operatorname{grad} \boldsymbol{u})^{\mathsf{T}}\right),$$

$$\text{material law: } \boldsymbol{\sigma} = \frac{\partial W}{\partial \boldsymbol{\varepsilon}}. \tag{2.1}$$

In order to derive the balance law for configurational forces, we follow the approach by [2] and compute the gradient of the strain energy W. The strain energy W depends on the strain $\boldsymbol{\varepsilon}$ and explicitly on the position $\boldsymbol{x}$, i.e $W = W(\boldsymbol{\varepsilon}; \boldsymbol{x})$. The computation of the gradient yields

$$\operatorname{grad} W = \frac{\partial W}{\partial \boldsymbol{\varepsilon}} : \operatorname{grad} \boldsymbol{\varepsilon} + \left.\frac{\partial W}{\partial \boldsymbol{x}}\right|_{\text{expl.}} = \boldsymbol{\sigma} : \operatorname{grad} \operatorname{grad} \boldsymbol{u} + \left.\frac{\partial W}{\partial \boldsymbol{x}}\right|_{\text{expl.}}, \tag{2.2}$$

where we have used the elastic material law, the symmetry of the stress $\boldsymbol{\sigma}$ and the kinematic equation, i.e. compatibility. The above equation can be written in index notation using Einstein's summation convention for repeated indices by

$$W_{,k} = \frac{\partial W}{\partial \varepsilon_{ij}} \varepsilon_{ij,k} + \left.\frac{\partial W}{\partial x_k}\right|_{\text{expl.}} = \sigma_{ij} u_{i,kj} + \left.\frac{\partial W}{\partial x_k}\right|_{\text{expl.}}. \tag{2.3}$$

The first term on the right-hand side can be rewritten by the use of the product rule and the equilibrium equation

$$\sigma_{ij} u_{i,kj} = \left(\sigma_{ij} u_{i,k}\right)_{,j} - \underbrace{\sigma_{ij,j}}_{-f_i} u_{i,k}. \tag{2.4}$$

Combining (2.3) and (2.4) yields the configurational force balance in symbolic notation

$$\text{div} \underbrace{\left(W\mathbf{1} - (\text{grad } \boldsymbol{u})^{\text{T}} \boldsymbol{\sigma}\right)}_{\boldsymbol{\Sigma}} \underbrace{- (\text{grad } \boldsymbol{u})^{\text{T}} \boldsymbol{f} - \left.\frac{\partial W}{\partial \boldsymbol{x}}\right|_{\text{expl.}}}_{\boldsymbol{g}} = \mathbf{0}, \tag{2.5}$$

where we observe a structure similar to the equilibrium condition. The divergence of the configurational stress tensor $\boldsymbol{\Sigma}$, sometimes also called Eshelby stress due to seminal works of Eshelby, is balanced by a configurational volume force $\boldsymbol{g}$. From (2.5) it is obvious that the Eshelby stress $\boldsymbol{\Sigma}$ is a conserved quantity, i.e. divergence free, if the physical volume forces vanish ($\boldsymbol{f} = \mathbf{0}$) and if the material is homogeneous and defect free $\left(\left.\frac{\partial W}{\partial \boldsymbol{x}}\right|_{\text{expl.}} = \mathbf{0}\right)$.

As discussed in [1, 19, 21, 27, 28] discrete solutions by standard displacement finite elements are continuous in the displacement $\boldsymbol{u}$ across element edges. Gradients, such as the strain, and thus the stresses are in general not continuous across element edges. In the sense of configurational mechanics the elements can be considered as small inhomogeneities, resulting in so-called 'spurious' or 'numerically' caused configurational forces. For details on the use of these spurious configurational forces in r- and h-adaptive schemes the reader is referred to [1, 18–21, 27–29] and the literature cited in there. These spurious configurational forces appear in regions with high strain gradients, see [20], and are therefore also an indicator for a necessary model refinement on the micro-level in multi-level methods. The next section is devoted to the basic concepts of a two-scale approach for the physical and configurational quantities.

2.3 Realisation of Micro–Macro Transition

In order to compute a residual on the macro-level in the Finite Element Method, the stress $\boldsymbol{\sigma}$ needs to be computed in every macroscopic (Gauß) integration point $\boldsymbol{x}$. On the micro-level different solution strategies can be applied, ranging from analytical homogenisation methods to numerical methods based on Boundary Element, Finite Element or Fast Fourier Methods. In general, we assume that a scale separation holds, i.e. the micro-level can be described by a representative volume element (RVE). The characteristic length of the RVE l must be small compared to the characteristic size of the macroscopic problem L ($l \ll L$). On the other hand the RVE must be large enough to cover the inhomogeneity on the micro-level. Following Hill's principle [9], the connection between the micro- and macro-level is given by averaging procedures, see also other textbooks on homogenisation and multi-scale approaches. Without claiming completeness we refer to [4, 22–24]. Thus, the macroscopic strain $\boldsymbol{\varepsilon}^*$ is given by

$$\boldsymbol{\varepsilon}^* = \langle \boldsymbol{\varepsilon} \rangle = \frac{1}{|V_{\mathrm{RVE}}|} \int_{V_{\mathrm{RVE}}} \boldsymbol{\varepsilon} \, \mathrm{d}V = \frac{1}{2|V_{\mathrm{RVE}}|} \int_{\partial V_{\mathrm{RVE}}} (\boldsymbol{u} \otimes \boldsymbol{n} + \boldsymbol{n} \otimes \boldsymbol{u}) \, \mathrm{d}A \,, \qquad (2.6)$$

where we have assumed that no displacement jumps occur on the micro-level, i.e. there are no cracks on the micro-level. In a similar way the macroscopic stress $\boldsymbol{\sigma}^*$ is defined by the volume average of the microscopic stress via

$$\boldsymbol{\sigma}^* = \langle \boldsymbol{\sigma} \rangle = \frac{1}{|V_{\mathrm{RVE}}|} \int_{V_{\mathrm{RVE}}} \boldsymbol{\sigma} \, \mathrm{d}V = \frac{1}{|V_{\mathrm{RVE}}|} \int_{\partial V_{\mathrm{RVE}}} \boldsymbol{t} \otimes \boldsymbol{x} \, \mathrm{d}A \,. \qquad (2.7)$$

In the above identity between the last two expressions, we have assumed that the volume forces $\boldsymbol{f}$ vanish on the micro-level. Using Hill's condition [9]

$$\langle \boldsymbol{\sigma} : \boldsymbol{\varepsilon} \rangle = \langle \boldsymbol{\sigma} \rangle : \langle \boldsymbol{\varepsilon} \rangle = \boldsymbol{\sigma}^* : \boldsymbol{\varepsilon}^* \qquad (2.8)$$

allows for the formulation of boundary conditions on the RVE. In order to do so, the microscopic displacement field is decomposed as

$$\boldsymbol{u}(\boldsymbol{x}) = \boldsymbol{u}^* + \mathrm{grad}\, \boldsymbol{u}^* \boldsymbol{x} + \boldsymbol{w}(\boldsymbol{x}) \,, \qquad (2.9)$$

where $\boldsymbol{u}^*$ and $\mathrm{grad}\, \boldsymbol{u}^*$ are the displacement and the displacement gradient at the macroscopic point $\boldsymbol{x}^*$ to which the RVE is attached. The fluctuations on the microscopic level are expressed by $\boldsymbol{w}$. Equation (2.6) implies that

$$\int_{\partial V_{\mathrm{RVE}}} \mathrm{grad}_{\mathrm{sym}}\, \boldsymbol{w} \, \mathrm{d}V = \frac{1}{2} \int_{\partial V_{\mathrm{RVE}}} (\boldsymbol{w} \otimes \boldsymbol{n} + \boldsymbol{n} \otimes \boldsymbol{w}) \, \mathrm{d}A = \boldsymbol{0} \,. \qquad (2.10)$$

The above relation is satisfied for prescribed linear displacements, i.e. $\boldsymbol{w} = \boldsymbol{0}$ on $\partial V_{\mathrm{RVE}}$ or periodic boundary conditions, where

$$\boldsymbol{w}^+ = \boldsymbol{w}^- \,. \qquad (2.11)$$

For a better understanding the two kinds of boundary conditions are sketched for a 2D situation in Fig. 2.1. By (2.11), a non-zero fluctuation at a corner of the RVE results in an identical fluctuation at all RVE corners. This induces a rigid body translation of the entire RVE which does not influence the strains and stresses inside the RVE. Without loss of generality, we can therefore assume vanishing fluctuations at the corners of the RVE. Using (2.9) one can decompose the gradient $\mathrm{grad}\, \boldsymbol{u}^*$ into a symmetric part related to the strain and skew symmetric part $\boldsymbol{R}^*$ related to the macroscopic rotation, such that

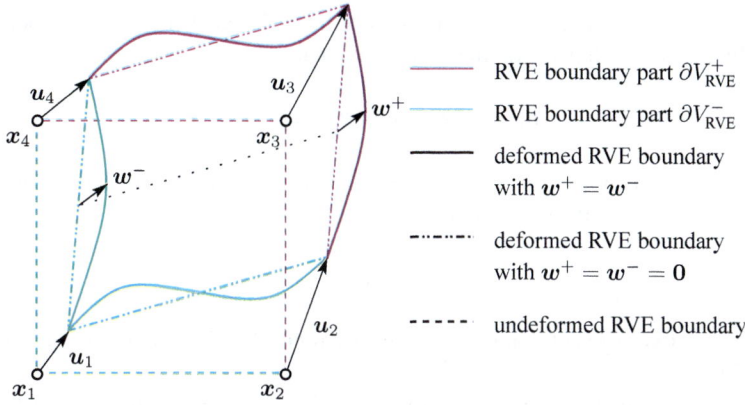

Fig. 2.1 Boundary conditions on the RVE: Linear displacements $w = 0$, and periodic boundary conditions: $w^+ = w^-$

$$u(x) = \left(u^* + R^*x\right) + \left(\varepsilon^*x + w\right), \tag{2.12}$$

where the first contribution is a rigid body motion (translation and rotation) of the RVE. Because of

$$\varepsilon = \mathrm{grad}_{\mathrm{sym}}\, u = \frac{1}{2}\left(R^* + R^{*\mathrm{T}}\right) + \frac{1}{2}\left(\varepsilon^* + \varepsilon^{*\mathrm{T}}\right) + \mathrm{grad}_{\mathrm{sym}}\, w = \varepsilon^* + \mathrm{grad}_{\mathrm{sym}}\, w \tag{2.13}$$

the macroscopic rigid body motion does not contribute to the microscopic strain and thus, the microscopic stress. This is the reason, why in classical two scale homogenisation schemes for infinitesimal strains the rigid body rotation R^* is not transferred between the scales. In the following we will neglect the macroscopic translation u^*, but pay special attention to the macroscopic rotation R^*. As an illustrative example, we consider the displacements of the corner points x_K of the RVE for two different scenarios. If the full displacement gradient is used, we get

$$u_{K|\,\mathrm{grad}\,u} = \mathrm{grad}\,u^* x_K, \tag{2.14}$$

while if the symmetric gradient, i.e. strain, is used, we obtain

$$u_{K|\varepsilon} = \varepsilon^* x_K. \tag{2.15}$$

The prescribed corner displacements differ by the macroscopic rigid body rotation, as

$$u_{K|\,\mathrm{grad}\,u} - u_{K|\varepsilon} = \left(\mathrm{grad}\,u^* - \varepsilon^*\right)x_K = R^* x_K. \tag{2.16}$$

If we suppose that

$$u_{\mathrm{grad}\,u} = \mathrm{grad}\,u^* x + w \tag{2.17}$$

is a solution to the boundary value problem in the RVE with the corner displacements given by (2.14) and a periodic fluctuation w, then

$$u_\varepsilon = \varepsilon^* x + w = u_{\text{grad } u} - R^* x \qquad (2.18)$$

is a solution to the boundary value problem that satisfies (2.15). However, both types of boundary conditions render identical strains, as

$$\varepsilon_{\text{grad } u} = \varepsilon^* + \text{grad}_{\text{sym}} \, w = \varepsilon_\varepsilon \, . \qquad (2.19)$$

Thus, the two types of boundary conditions render the same macroscopic stresses via (2.7).

In the following we will pay attention to the Eshelby stress Σ. In a straightforward way, one defines the macroscopic Eshelby stress by the same averaging procedure as for the physical stress, resulting in

$$\Sigma^* = \langle \Sigma \rangle = \frac{1}{|V_{\text{RVE}}|} \int_{V_{\text{RVE}}} \Sigma \, dV = \frac{1}{|V_{\text{RVE}}|} \int_{V_{\text{RVE}}} \left(W \mathbf{1} - (\text{grad } u)^{\text{T}} \sigma \right) dV \, . \quad (2.20)$$

It is also possible to define an Eshelby stress tensor based on homogenised macroscopic quantities. However, an Eshelby stress tensor computed in this manner is not consistent with the Hill condition. A detailed discussion of different possible definitions of the macroscopic Eshelby stresses is omitted here. The interested reader is referred to [11, 25, 26]. Using definition (2.20) one has to be careful regarding the macroscopic rotation R^*. While the strain ε and the stress σ and consequently the strain energy W are not affected by the two different boundary conditions for the RVE, the displacement gradients grad $u_{\text{grad } u}$ and grad u_ε differ by the macroscopic rotation R^*, as can be seen from

$$\text{grad } u_{\text{grad } u} - \text{grad } u_\varepsilon = \left(\text{grad } u^* + \text{grad } w \right) - \left(\varepsilon^* - \text{grad } w \right) = R^* \, . \qquad (2.21)$$

Thus, the Eshelby stresses $\Sigma_{\text{grad } u}$ and Σ_ε differ in the RVE resulting in different homogenised macroscopic Eshelby stresses. On the micro-level both Eshelby stresses can be used to evaluate for example defects on the micro-level. But in the homogenisation procedure only

$$\Sigma^*_{\text{grad } u} = \langle \Sigma_{\text{grad } u} \rangle$$

is physically meaningful. This suggests that in the configurational context the use of the displacement gradient boundary conditions according to grad u^* is to be preferred. If strain boundary conditions according to ε^* are used the direct homogenisation leads to unphysical results, but the homogenised Eshelby stress can be corrected:

Fig. 2.2 Multi-scale approach for configurational forces based on gradient approach

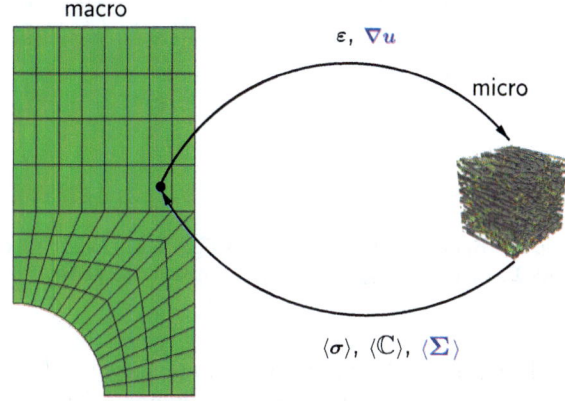

Fig. 2.3 Crack specimen with homogeneous and inhomogeneous RVE. Data used: $a = 10$ cm, $u_0 = 1$ mm, $E_0 = 70$ GPa, $v_0 = 0.3$, $E_1 = 10$ GPa, $v_1 = 0.3$, $E_2 = 300$ GPa, $v_2 = 0.3$

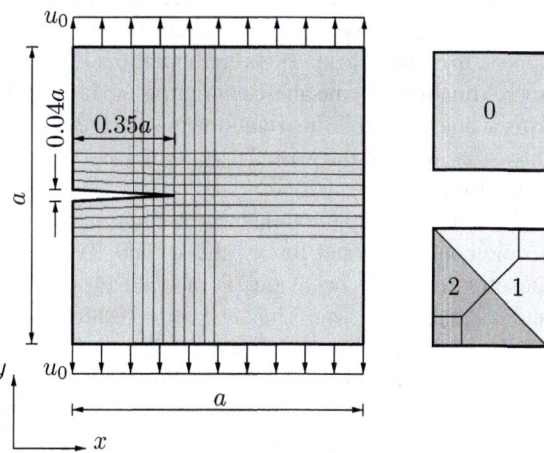

$$\boldsymbol{\Sigma}^*_{\text{grad }u} = \langle \boldsymbol{\Sigma}_{\text{grad }u} \rangle = \frac{1}{|V_{\text{RVE}}|} \int_{V_{\text{RVE}}} \left(W\mathbf{1} - (\text{grad } \boldsymbol{u}_{\text{grad }u})^{\text{T}} \boldsymbol{\sigma} \right) dV$$

$$= \frac{1}{|V_{\text{RVE}}|} \left[\int_{V_{\text{RVE}}} \left(W\mathbf{1} - (\text{grad } \boldsymbol{u}_{\varepsilon})^{\text{T}} \boldsymbol{\sigma} \right) dV \right] - \frac{1}{|V_{\text{RVE}}|} \int_{\text{RVE}} \boldsymbol{R}^{*\text{T}} \boldsymbol{\sigma} \, dV \quad (2.22)$$

$$= \langle \boldsymbol{\Sigma}_{\varepsilon} \rangle - \boldsymbol{R}^{*\text{T}} \langle \boldsymbol{\sigma} \rangle = \boldsymbol{\Sigma}^*_{\varepsilon} - \boldsymbol{R}^{*\text{T}} \boldsymbol{\sigma}^* .$$

The correction term $\boldsymbol{R}^{*\text{T}} \boldsymbol{\sigma}^*$ allows the use of strain boundary conditions on the RVE if these are preferred due to existing numerical two scale schemes.

The entire multi-scale approach in conjunction with the gradient based approach for the configurational forces is summarised in Fig. 2.2.

To illustrate the effects of the use of different RVE boundary conditions, Fig. 2.3 shows a two-scale setup, where macroscopically a situation with a crack under a mode I load is chosen. The configurational forces that result from a homogeneous

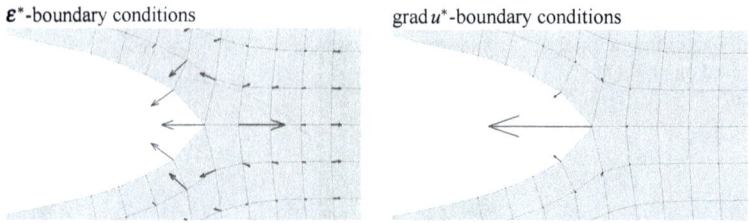

Fig. 2.4 Homogeneous RVE: nodal configurational forces in the vicinity of the crack tip computed with $\boldsymbol{\Sigma}_{\boldsymbol{\varepsilon}}$ (left) and $\boldsymbol{\Sigma}_{\mathrm{grad}\,\boldsymbol{u}}$ (right). Deformation enlarged by a factor of 10

RVE, top right sketch in Fig. 2.3, are shown in Fig. 2.4. It is expected that only at the crack tip a significant configurational force appears. This is the case if the Eshelby stress based on the displacement gradient boundary condition $\boldsymbol{\Sigma}_{\mathrm{grad}\,\boldsymbol{u}}$ is used. For $\boldsymbol{\Sigma}_{\boldsymbol{\varepsilon}}$ unphysical configurational forces in the close vicinity of the crack tip appear. In contrast to spurious configurational forces that appear due to the numerical approximation scheme, these configurational forces do not vanish for refined meshes. This is due to the missing rotational part in the computation of $\boldsymbol{\Sigma}_{\boldsymbol{\varepsilon}}$. Having identified this issue, we will only use $\boldsymbol{\Sigma}_{\mathrm{grad}\,\boldsymbol{u}}$ or $\boldsymbol{\Sigma}_{\boldsymbol{\varepsilon}}$ in combination with the correction (2.23) in the following.

In order to study the influence of inhomogeneous microstructures on the macroscopic configurational forces, the second RVE in Fig. 2.3 is considered. Here an anisotropic elastic behaviour is induced by the RVE composed of two materials and a slanted interface. The anisotropy results in a non-symmetric deformation, as can be seen from Fig. 2.5. It can be seen that the vertical mode I type load results in significant tilt in the horizontal direction due to the anisotropy induced by the RVE. In addition the configurational force deviates from the horizontal direction indicating a kinked crack propagation. It should be noted, that due to their definition, configurational forces point in the direction of an energy increase. Thus, if one

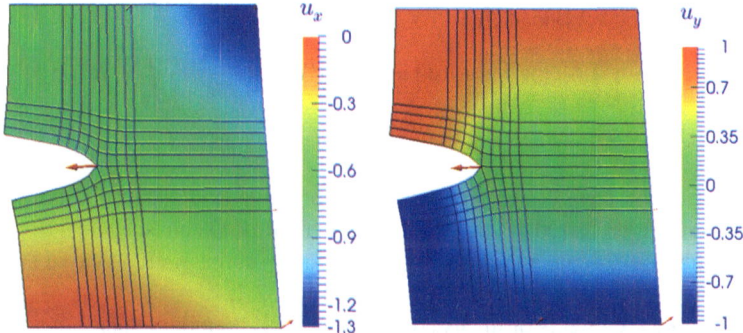

Fig. 2.5 Inhomogeneous RVE: deformation and configurational force at the crack tip. Deformation enlarged by a factor of 10

assumes crack propagation in the direction of an energy release, a crack will propagate in the negative direction of a configurational force. For a discussion on the issue of crack propagation by con-figu-rational forces the reader is referred to [5, 6, 16, 17, 19] and the literature cited in there.

2.4 Numerical Evaluation of Developed Indicators

In this section the application of configurational forces as an indicator for a refined two scale homogenisation is demonstrated. As an introductory remark Fig. 2.6 indicates different sources of configurational forces. Figure 2.6a, b shows configurational forces originating from inhomogeneities, namely a crack in the case of (a), and an interface in (b). The last example in Fig. 2.6c is defect-free. By theoretical considerations no configurational forces are expected in the interior. Due to the rather coarse numerical discretisation and the non-continuous stress and strain interpolation across the element interfaces, configurational forces appear. They become larger in regions with higher stress and strain gradients. In h-adaptive schemes, such as reported in [20], the mesh is refined in such regions. However, in a multi-scale approach areas with higher stress and strain gradients also require a more refined micromechanical modelling. Thus, in the following examples a strategy is presented, that allows for marking areas that require a highly resolved micro-scale model based on macroscopic configurational forces.

For an isotropic setup only the absolute value of a discrete configurational force G at node x_I in the discretised domain Ω_h is considered, i.e.

$$\| G(x_I) \| = \sqrt{G_1^2 + G_2^2 + G_3^2} \quad \text{for} \quad x_I \in \Omega_h . \tag{2.23}$$

The two step scheme works as follows:

(1) A filtering step eliminates large configurational forces that are physically relevant and appear for example at crack tips, interfaces or boundaries. A factor f is introduced. The set of filtered nodes x_f is given by

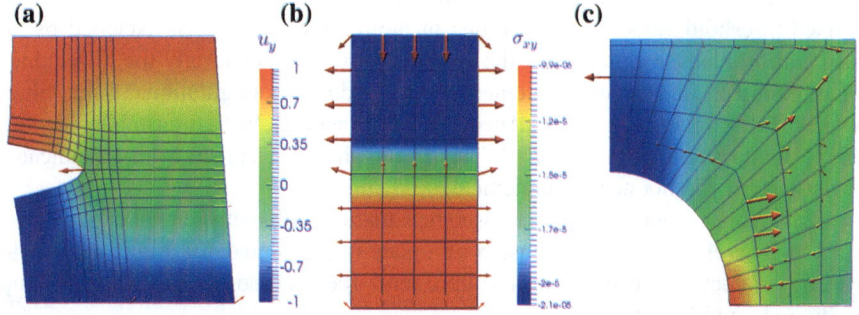

Fig. 2.6 Sources of configurational forces: **a** crack tip, **b** interfaces, **c** numerical discretisation

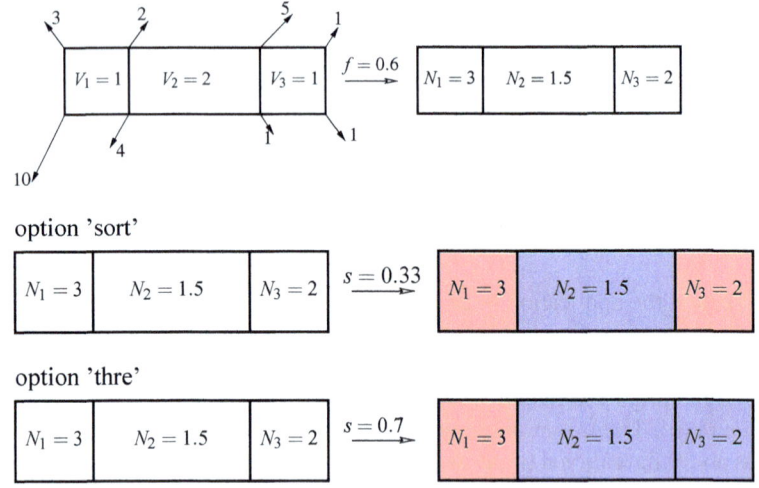

Fig. 2.7 Marking strategy based on filtered configurational forces: comparison of sorting (sort) and threshold (thre) procedure

$$x_f = \{x_I \in \Omega_h| \parallel G(x_I) \parallel < f \parallel G \parallel_{max}\} , \qquad (2.24)$$

with the maximal absolute value

$$\parallel G \parallel_{max} = \max_{x_I \in \Omega_h} \parallel G(x_I) \parallel . \qquad (2.25)$$

(2) In a second step, based on an element indicator, defined by

$$N_e = \frac{1}{\parallel x_f \parallel V_e} \sum \parallel G(x_f) \parallel \qquad \text{for } x_f \in \Omega_e , \qquad (2.26)$$

in which $\parallel x_f \parallel$ is the number of filtered nodes on the element, elements are marked for a refined micromechanical modelling. The marking is performed either based on a percentage of elements from a sorted list of element indicators, or based on a fixed threshold value. The main idea of the marking strategy is sketched by an artificial problem in Fig. 2.7. In the sorting procedure (sort), only the lower third of all elements are selected as elements with a reduced configurational force (blue) while the remaining elements (red) need a RVE refinement. In the threshold procedure (thre) the threshold is set at $0.7 \cdot 3$. This means that all elements with an indicator lower than 2.1 do not need a refinement.

The strategy is applied to a macroscopic situation with realistic application background. Therefore a compact tension of a plate is considered. To mimic the mould casting production process of a short fibre reinforced composite, a three layer setup is studied, see Fig. 2.8.

Fig. 2.8 Three layer setup of
a short fibre reinforced plate
in compact tension loading

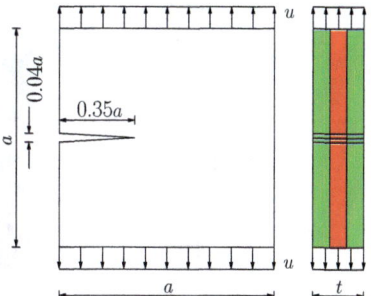

Fig. 2.9 Configurational
forces: **a** in the centre, **b** in
the interface

(a)   **(b)**

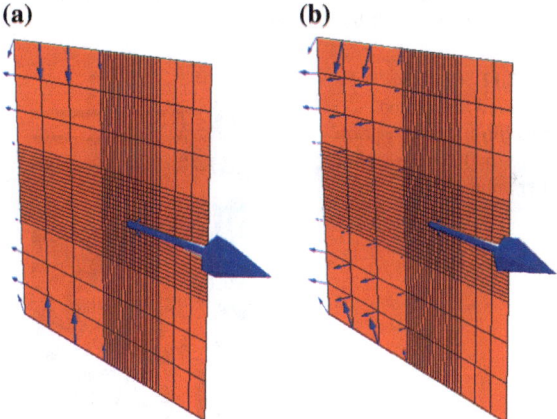

The different layers are made up of composite materials with different fibre structures. The elastic responses of the different fibre orientations of the three layers were
computed using the micromechanical software GeoDict [3], which was also used
to generate two different RVEs that are used in the micromechanical computations.
The effective elasticity tensors and the RVEs are assigned to the two external layers
and the internal layer, respectively. The resulting configurational forces are depicted
in Fig. 2.9.

Subplot Fig. 2.9a shows the configurational forces in the central layer of elements.
As expected large configurational forces appear at the crack tip. These configurational
forces are of physical origin and, as discussed above, can be regarded as a crack
driving force. In addition, configurational forces appear at the boundary. They can be
regarded as reaction forces, that appear as reaction to fixing the geometry. In Fig. 2.9b
physically motivated configurational forces appear in the interface. The interface
between the layers with different elastic properties represents an inhomogeneity or
a defect, thus configurational forces appear.

In the first example (Figs. 2.10 and 2.11), the filter parameter is set to $f = 1$. This
leads to an unfiltered indicator in which physical as well as numerical configurational
forces are taken into account. In Fig. 2.10 the sorting option is then used to identify
regions of refined RVE modelling, which are represented as transparent elements

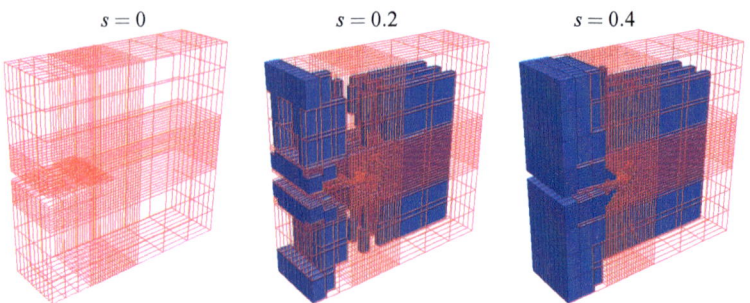

Fig. 2.10 Elements identified for refined modelling: sorting option with $s = 0$ (non marked), $s = 0.2$, and $s = 0.4$; filter value $f = 1$

Fig. 2.11 Elements identified for refined modelling: sorting option with $s = 0.2$ and $s = 0.4$; filter value $f = 1$

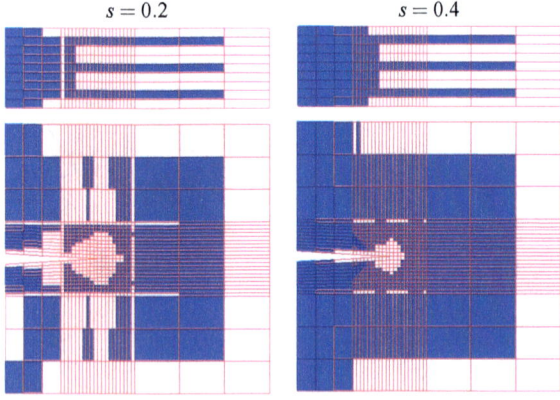

(high indicator value) while the blue regions do not need a refined treatment (low indicator value). Because the 3D view is not very easy to comprehend, 2D views are provided in Fig. 2.11.

As can be seen from Figs. 2.10 and 2.11 the sorting method is capable of capturing the vicinity of the crack tip in all three layers as well as the regions of the material interface as relevant region for a refined modelling. Since the filtering is omitted in this case, physical and numerical configurational forces determine the refinement regions of the mesh.

The marking strategy based on a threshold value is reported in Fig. 2.12. It is observed, that for the chosen threshold values, a very small region is detected for refinement. In Fig. 2.13 a larger region for refinement is reported. This is achieved by the same threshold values of $s = 0.2$, $s = 0.4$, and $s = 0.6$, but the filtering step is modified by changing the value of f from 10^{-4} in Fig. 2.12 to 10^{-5} in Fig. 2.13. By this means, only the smallest configurational forces, which generally have a numerical character, are considered for the indicator. It should also be noted that in this example the unfiltered option seems to be stronger influenced by the presence of the interfaces between the layers, while the filtering method is more homogeneous throughout the thickness of the specimen.

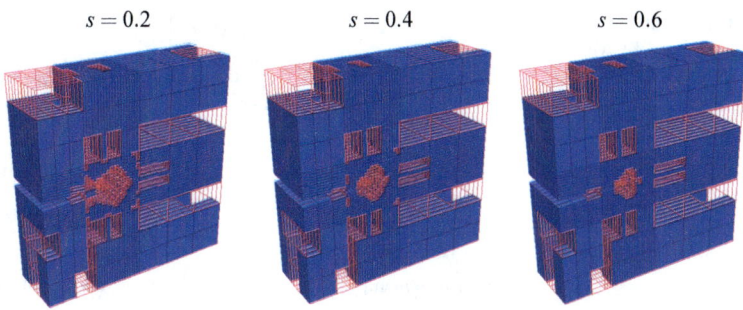

Fig. 2.12 Elements identified for refined modelling: threshold option with $s = 0.2$, $s = 0.4$, and $s = 0.6$; filter value $f = 10^{-4}$

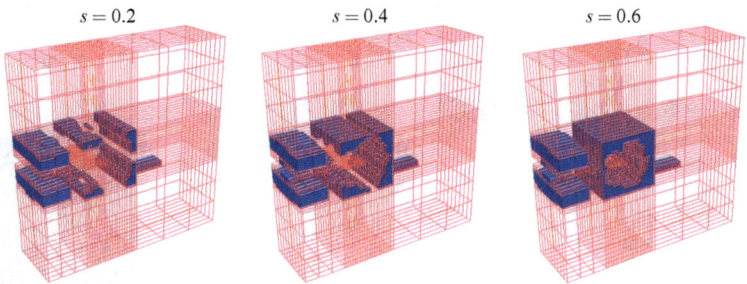

Fig. 2.13 Elements identified for refined modelling: threshold option with $s = 0.2$, $s = 0.4$, and $s = 0.6$; filter value $f = 10^{-5}$

2.5 Scalability

As the multi-scale computations are extremely time consuming, due to the fact that each macroscopic finite element integration point requires the solution of a complete boundary value problem, parallel computations are of importance. In order to achieve scalability on multi-processor machines the finite element programme FEAP (Finite Element Analysis Program) from UC Berkeley is used. It has a build in (FE)^2 implementation with parallel execution of the RVEs. However, to enable the coupling to other solvers on the micro-level of the RVE, a PYTHON interface is used. This allows for the coupling of Fast Fourier Transform based solvers such as FeelMath, boundary integral based solvers, and advanced analytical homogenisation techniques in a very flexible way. The implementation of the coupling strategy described above is depicted in Fig. 2.14, where FEAP itself is used as the solver on the micro-level. It is emphasised here again that the PYTHON script "microproblem.py" can incorporate any other micromechanical solver. As a benchmark to check the scalability a macroscopic tension specimen is simulated. Details on the geometry, discretisation, RVE, and material data are given in Fig. 2.15. The scalability with an increasing number of processors to calculate the RVEs is presented in Table 2.1. It is obvious

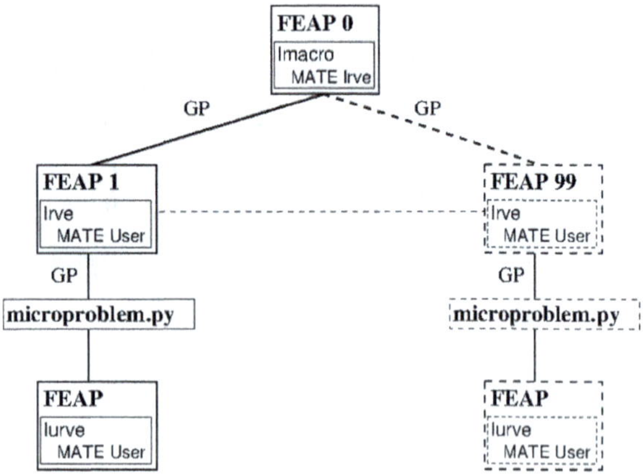

Fig. 2.14 Parallel implementation in FEAP using a PYTHON based coupling interface

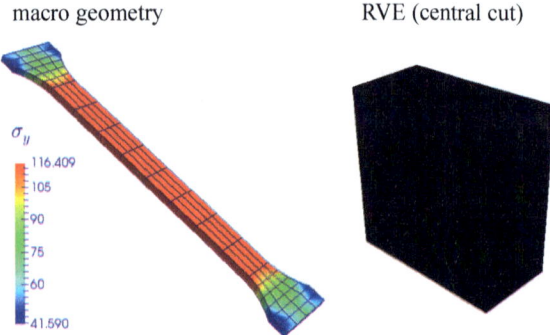

Fig. 2.15 Parallel (FE)^2 computation using FEAP: specimen length $l = 115$ mm, displacement load $u = 1$ mm, macro-level: 80 elements, 640 Gauß points, micro-level: 216 elements, Young's moduli $E_1 = 10$ GPa (blue), $E_2 = 70$ GPa (red), Poisson's ratios: $|v_1 = v_2 = 0.3$

macro geometry   RVE (central cut)

Table 2.1 Scalability of benchmark problem, computation time in seconds: Comparison of PYTHON interface with FEAP internal (FE)^2 implementation

CPUs	PYTHON	FEAP (FE)^2
2	254.98	30.60
4	132.10	18.73
8	67.43	12.46
16	37.18	5.81
32	22.04	3.75
64	18.09	2.47

from Table 2.1 that both implementations, the PYTHON based interface and the FEAP internal (FE)^2, scale with the number of CPUs. The PYTHON based interface relies on the execution of a PYTHON script started from within the running FEAP program and all the communication is done via files. Thus, its performance is slower than the direct use of the MPI communication environment running in the compiled FEAP internal (FE)^2 approach. But, as discussed above, the PYTHON interface can readily be used to couple other existing micro-solvers without the need to couple source codes by the MPI broadcast environment, which can be very cumbersome.

Acknowledgements The authors thank J. Hebel and Md. Khalaquzzaman for the fruitful discussions and productive cooperation within the project MUlti-scale SImulation of COmposites (MUSIKO).

References

1. Braun, M.: Configurational forces induced by finite-element discretization. Proc. Estonian Acad. Sci. Phys. Math. **46**(1/2), 24–31 (1997)
2. Eshelby, J.D.: Energy relations and the energy-momentum tensor in continuum mechanics, pp. 77–115. In: Kanninen [9] (1970)
3. GeoDict.: www.geodict.com. Accessed 16 Jan 2019
4. Gross, D., Seelig, T.: Fracture Mechanics: With an Introduction to Micromechanics. Springer (2017)
5. Gürses, E., Miehe, C.: A computational framework of three-dimensional configurational-force-driven brittle crack propagation. Comput. Method Appl. Mech. Eng. **198**(15–16), 1413–1428 (2009)
6. Gurtin, M., Podio-Guidugli, P.: Configurational forces and the basic laws for crack propagation. JMPS **44**(6), 905–927 (1996)
7. Gurtin, M.E.: Configurational Forces as Basic Concept of Continuum Physics. Springer, Berlin, New York, Heidelberg (2000)
8. Heintz, P., Larsson, F., Hansbo, P., Runnesson, K.: Adaptive strategies and error control for computing materialforces in fracture mechanics. Int. J. Numer. Method Eng. **60**, 1287–1299 (2004)
9. Hill, R.: On constitutive macro-variables for heterogeneous solids at finite strain. Proc. R. Soc. Lond. A **326**(1565), 131–147 (1972)
10. Kanninen, M.F. (ed.): Inelastic Behaviour of Solids. McGraw Hill, New York (1970)
11. Khalaquzzaman, M., Xu, B.X., Ricker, S., Müller, R.: Computational homogenization of piezoelectric materials using FE^2 to determine configurational forces. Tech. Mech. **32**(1), 21–37 (2012)
12. Kienzler, R., Herrmann, G.: Mechanics in Material Space. Springer, New York, Berlin, Heidelberg (2000)
13. Kuhn, C., Müller, R., Klassen, M., Gross, D.: Numerical homogenization of the Eshelby tensor at small strains. Math. Mech. Solids (2017). https://doi.org/10.1177/1081286517724607
14. Maugin, G.A.: Material Inhomogeneities in Elasticity. Chapman & Hall, London, Glasgow, New York, Tokyo, Melbourne, Madras (1993)
15. Maugin, G.A.: Configurational Forces—Thermomechanics, Physics, Mathematics, and Numerics. CRC Press, Boca Raton, London, New York (2011)
16. Miehe, C., Gürses, E.: A robust algorithm for configurational force driven brittle crack propagation with r-adaptive mesh alignment. IJNME **72**, 127–155 (2007)
17. Miehe, C., Gürses, E., Birkle, M.: A computational framework of configurational-force-driven brittle fracture based on incremental energy minimization. Int. J. Fract. **145**(4), 245–259 (2007)
18. Molser, J., Ortiz, M.: On the numerical implementation of variational arbitrary Lagrangian–Eulerian (VALE) formulations. Int. J. Numer. Method Eng. **67**(9), 1272–1289 (2006)
19. Mueller, R., Maugin, G.: On material forces and finite element discretizations. Comput. Mech. **29**(1), 52–60 (2002)
20. Mueller, R., Gross, D., Maugin, G.: Use of material forces in adaptive finite element methods. Comput. Mech. **33**, 421–434 (2004)
21. Mueller, R., Kolling, S., Gross, D.: On configurational forces in the context of the Finite Element Method. Int. J. Numer. Method Eng. **61**(1), 1–21 (2004)
22. Mura, T.: Micromechanics of Defects in Solids. Martinus Nijhoff Publishers (1987)

23. Nemat-Nasser, S., Hori, M.: Micromechanics: Overall Properties of Heterogeneous Materials. North Holland, Amsterdam, London, New York, Tokyo (1993)
24. Qu, J., Cherkaoui, M.: Fundamentals of Micromechanics of Solids. Wiley (2007)
25. Ricker, S., Mergheim, J., Steinmann, P.: On the multiscale computation of defect driving forces. Int. J. Multiscale Comput. Eng. **7**(5) (2009). https://doi.org/10.1615/IntJMultCompEng.v7.i5.70
26. Ricker, S., Mergheim, J., Steinmann, P., Müller, R.: A comparison of different approaches in the multi-scale computation of configurational forces. Int. J. Fract. **166**, 203–214 (2010)
27. Steinmann, P.: Application of material forces to hyperelastic fracture mechanics. I. Continuum mechanical setting. Int. J. Solids Struct. **37**(48–50), 7371–7391 (2000)
28. Steinmann, P., Ackermann, D., Barth, F.J.: Application of material forces to hyperelastic fracture mechanics. II. Computational setting. Int. J. Solids Struct. **38**(32–33), 5509–5526 (2001)
29. Thoutireddy, P., Ortiz, M.: A variational r-adaption and shape-optimization method for finite deformation elasticity. Int. J. Numer. Method Eng. **53**, 1557–1574 (2002)

Chapter 3
Modelling of Geometrical Microstructures and Mechanical Behaviour of Constituents

Heiko Andrä, Dascha Dobrovolskij, Katja Schladitz, Sarah Staub and Ralf Müller

3.1 Analysis of Fibre Orientation for Glass Fibre Reinforced Polymers Based on µCT Scans

3.1.1 Sample Preparation and Analysis of Fibre Direction

Throughout this chapter, a 2 mm thick polybutylene terephthalate (PBT) plate reinforced with 20 weight percent glass fibres is considered. First, the material is spatially imaged by micro-computed X-ray tomography (µCT), in order to determine analytically essential micro-structure features. Measuring the fibre orientation requires µCT with nominal resolutions in the range well below 10 µm. To achieve this with a standard laboratory CT setup, samples of a few millimetre diameter have to be extracted from the plate. These samples are extracted from the plate according to the scheme shown in Fig. 3.1. To choose five samples is a compromise between the effort for imaging and analysing on the one hand and capturing the systematic microstructural

H. Andrä (✉) · D. Dobrovolskij · K. Schladitz · S. Staub
Fraunhofer Institute for Industrial Mathematics, 67663 Kaiserslautern, Germany
e-mail: heiko.andrae@itwm.fraunhofer.de

D. Dobrovolskij
e-mail: dascha.dobrovolskij@itwm.fraunhofer.de

K. Schladitz
e-mail: katja.schladitz@itwm.fraunhofer.de

S. Staub
e-mail: sarah.staub@itwm.fraunhofer.de

D. Dobrovolskij
University of Applied Sciences, 64295 Darmstadt, Germany

R. Müller
University of Kaiserslautern, 67653 Kaiserslautern, Germany
e-mail: ram@rhrk.uni-kl.de

© Springer-Verlag GmbH Germany, part of Springer Nature 2019   31
S. Diebels and S. Rjasanow (eds.), *Multi-scale Simulation of Composite Materials*,
Mathematical Engineering, https://doi.org/10.1007/978-3-662-57957-2_3

Fig. 3.1 Moulded glass fibre reinforced PBT plate with marked specimen positions

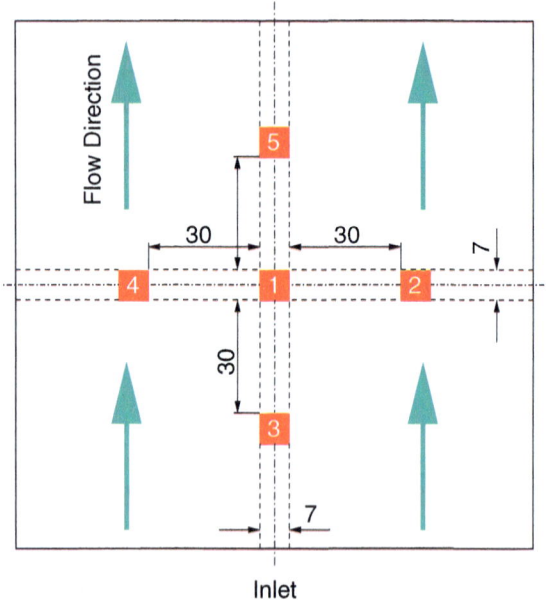

Inlet

differences expected at different positions w.r.t. the inlet or the flow front, respectively. The positions are thus chosen according to the expected behaviour of the latter. The five taken cuboidal samples of base edge length 2 mm are imaged using ITWM's CT device at a tube voltage of 160 kV and using a flat bed PerkinElmer and a Thales detector, respectively. A voxel edge length of 1.2–1.3 µm is chosen, such that the fibre diameter is resolved by approximately ten voxels. The resulting 2D projection images are reconstructed using weighted filtered back projection, and the microstructures are analysed based on the obtained three dimensional image data.

The analysis focuses on the fibre orientation, in particular in order to determine the thickness of the different layers, which are typical for injection moulded GFRP plates, and the main fibre orientation in each layer. In the following, the theoretical background of the applied fibre orientation analysis method is summarised. Subsequently, the pre- and post-processing is described. The analysis results for the five samples are finally presented in Sect. 3.1.2.

The fibre directions in the µCT images are analysed by means of the "SubField-FiberDirection" function of ITWM's software tool MAVI [4]. This function computes, based on the local grey values, for each voxel the local fibre orientation. From these local orientations, restricted to the fibre component, local orientation tensors are derived. The following two paragraphs summarise very shortly the mathematical basis of the local orientation measurement. The fibre component is interpreted mathematically as a random closed set Φ in the three-dimensional Euclidean space, see e.g. [15, 18, 20]. The fibre direction distribution in a typical point of this set corresponds to a measure in the space of non-oriented directions

$$R(A) = \frac{1}{2\pi V(W)} \mathbb{E} \int_{W \cap \Phi} I_A(v(x)) \, dx, \tag{3.1}$$

where $A \subseteq S_+^2$ denotes a measurable set of non-oriented directions, $v(x)$ the direction of the fibre system at location $x \in W$, and $W \subseteq \mathbb{R}^3$ the considered region. The expectation $\mathbb{E}$ is computed with respect to the distribution law of the random set Φ. The function I_A is an indicator for the set A, i.e. $I_A(v) = 1$ if $v \in A$ and otherwise 0.

The local fibre direction in each voxel is derived from the Hessian matrix—the matrix of second order grey value partial derivatives. In order to compute it, the original image $f : W \cap \mathbb{L} \mapsto \{0, \ldots, 255\}$ is first smoothed by a Gaussian filter g_σ of size σ. Here $\mathbb{L} = s\mathbb{Z}^3$, $s \in \mathbb{R}$ denotes a 3D orthogonal isotropic lattice. Then the Hessian matrix $H(x)$ is approximated by finite difference quotients:

$$H_{ij}(\mathbf{x}) = \left(\frac{\partial^2}{\partial x_i \partial x_j}\right)(f * g_\sigma)(\mathbf{x}), \quad i, j = 1, 2, 3, \ \mathbf{x} \in \mathbb{L}. \tag{3.2}$$

Let $|\lambda_1| \leq |\lambda_2| \leq |\lambda_3|$ be the eigenvalues of the Hessian H, ordered with respect to magnitude. Then the local direction of the fibre system Φ in x is given as the direction corresponding to the smallest (in magnitude) eigenvalue $|\lambda_1|$ as glass fibres appear bright compared to the polymer matrix surrounding them, see [3, 22]. Roughly, the idea behind this approach is a local approximation of the fibre by a cylinder of thickness 2σ in each voxel. The local fibre direction is then associated with the minimal curvature of the grey value formation, the spatial direction in which the least change is observed.

Initially, the method yields a local orientation in each voxel. In a second step, the result is masked with a segmentation of the fibre system. Note that this means just, that the fibre system has to be separated from the polymer matrix. The segmentation of individual fibres in the 3D image is however not needed.

Given the distribution R of the fibre orientation in the typical point of the fibre system, the second order orientation tensor a is derived as the outer product of the components v_1, v_2, v_3 of the orientation vector averaged with respect to R

$$a_{ij} = \int v_i v_j R(dv), \quad i, j = 1, 2, 3. \tag{3.3}$$

Thus, based on the image data, the orientation tensor is computed by averaging the voxel-wise product over defined sub-volumes $W_0 \subseteq W$

$$\hat{a}_{ij} = \sum_{x \in W_0 \cap \mathbb{L} \cap \Phi} v_i(\mathbf{x}) v_j(\mathbf{x}), \tag{3.4}$$

see [21]. The main fibre direction $\bar{v}$ in sub-volume W is obtained as the eigenvector associated to the largest eigenvalue of the tensor a. Of course, this main fibre direction is meaningful only if the distribution R has a cluster-like shape [2].

Usually, the fibre system is segmented from the 3D image by a simple global thresholding. This requires to remove global grey value fluctuations within the image—a task easily achieved by so-called shading correction: The image is smoothed extensively and subsequently subtracted from the original. Afterwards, a global grey value threshold suffices to separate the bright fibres from the dark matrix. MAVI's "SubFieldFiberDirection" suggests to apply the threshold according to Otsu [17] multiplied by 1.25 to avoid erroneous orientation information at the fibre edges.

In this study, this usual procedure turned out to distort the strongly anisotropic fibre orientation distributions within the layers towards isotropy. Therefore, to ensure that exclusively voxels from the fibre cores contribute, the fibre system is segmented based on Frangi's vesselness index [3]. Again, the eigenvalues of the Hessian matrix are exploited, this time to gain local structure shape information. The structure is locally fibrous if and only if there are one small and two large eigenvalues (in magnitude). Frangi's index is designed for detecting bright fibres on dark background. Thus it is non-zero only if $\lambda_2, \lambda_3 < 0$. In this case it is computed as

$$\left(1 - \exp\left(-2\lambda_2^2/\lambda_3^2\right)\right) \exp\left(-2\lambda_1^2/|\lambda_2\lambda_3|\right)\left(1 - \exp\left(-2(\lambda_1^2 + \lambda_2^2 + \lambda_3^2)/c^2\right)\right)$$

with $c^2 = (\max_{\mathbf{x} \in W \cap \mathbb{L}} \|H(\mathbf{x})\|_F)^2 = \max_{\mathbf{x} \in W \cap \mathbb{L}}(\lambda_1(\mathbf{x})^2 + \lambda_2(\mathbf{x})^2 + \lambda_3(\mathbf{x})^2)$. This index is now computed and used to derive a more precise segmentation of the fibre system. That is, the global grey value threshold is applied on the image holding in each voxel the local Frangi's index.

3.1.2 Results

As a preliminary step, the three-dimensional images are rotated such that the imaged cuboidal sample is oriented parallel to the coordinate axes. More precisely, the x-direction corresponds to the injection direction, y to the cross-flow direction and z to the direction in thickness of the plate.

Second, the layers are approximated. To this end, the images are first denoised by a $5 \times 5 \times 5$ pixel median filter. Then, the grey values are averaged along rays in y-direction. This results in distinct bright spots the x–z-plane in the central region, where high grey values are summed along fibres, see Fig. 3.2, right. The layers are finally deduced from the curves of the row-wise grey value maxima: The global maximum of the B-spline smoothed graph indicates the centre. The lower and the upper bounds are derived from the two local minima closest to the global maximum. More precisely, we chose the argument of the higher of these two minima to define the half-width of the symmetric interval, see Fig. 3.2, left.

This yields the thickness values as reported in Table 3.1. The orientation tensor for each layer is derived by averaging over the complete layer. A volume rendering of

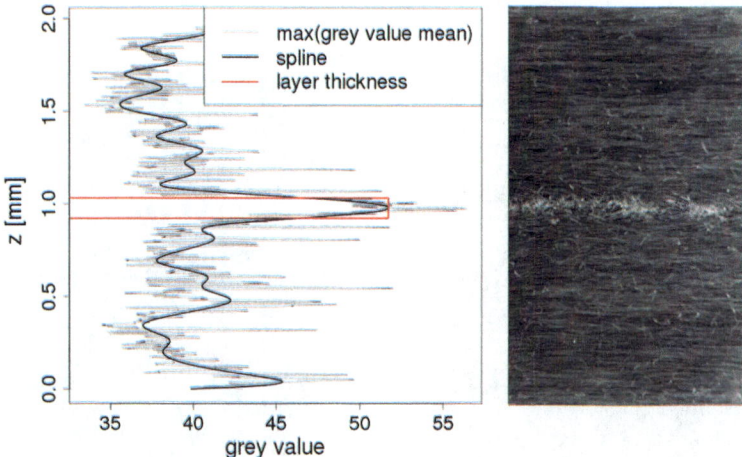

Fig. 3.2 Grey value means for determining the misoriented central layer, illustrated for sample 5. Left: row-wise maxima of averaged grey values. Right: grey value means along stripes in y-direction

the fibre system contributing to the orientation results is shown in Fig. 3.3 illustrating the orientation tensor results for the total thickness of the plate.

Subsequently, the local fibre orientation is analysed as described in Sect. 3.1.1 above. This yields the second order orientation tensors in the three layers, see Table 3.1. As expected, the fibres are mainly oriented along the x-axis (flow direction) in the outer layers, whereas the fibres are oriented orthogonal to the flow direction in the central layer.

The fibre orientation in the misoriented layers differs with respect to proportions of fibres oriented in x- and y-directions (see Table 3.1 columns a_{xx} and a_{yy}). Figures 3.4 and 3.5 contain slices through the upper and the misoriented central layers of all five samples. Both these views as well as the quantitative results from Table 3.1 confirm the expected curved flow front [14]. The main fibre orientation in the central layers in the outer plate regions are tilted towards the plate edges in x-direction. This observation is in perfect accordance with the expected faster fibre transport at the outer regions compared to the inner plate part.

To summarise, in all three layers the fibres are almost oriented in-plane, i.e. the thickness component z of the orientation tensor is small compared to the entries in x- and y-direction. For all samples the flow direction is the governing direction in the outer layers and the fibres are re-oriented in the central layer. Thus, virtually generated volume elements should take into account the observed multi-layer composition in order to represent the microstructure appropriately.

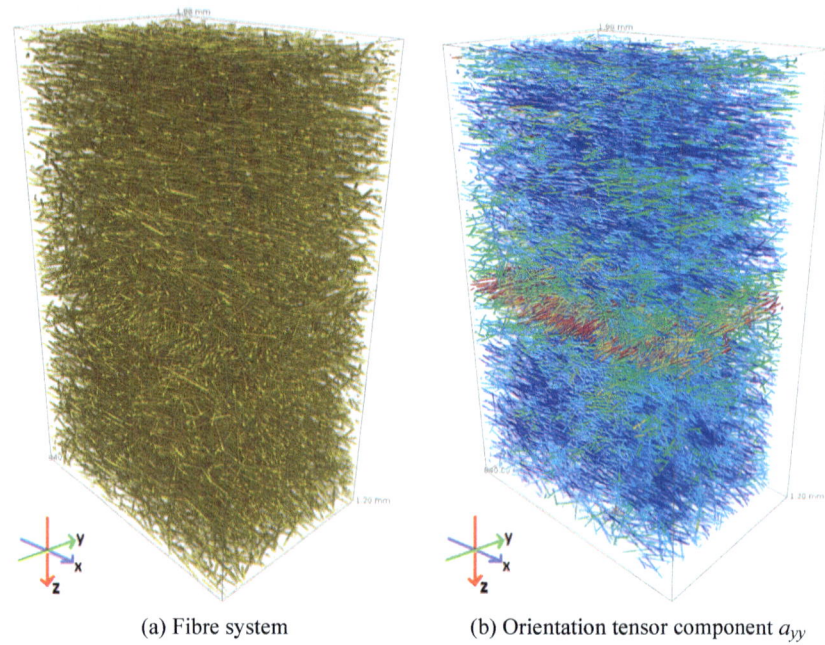

(a) Fibre system   (b) Orientation tensor component a_{yy}

Fig. 3.3 Volume renderings of sample 5. Visualised are approximately 1.2 mm × 0.9 mm × 2.0 mm out of the totally analysed approximately 1.8 mm × 1.1 mm × 2.0 mm. Fibre system as obtained by thresholding the image holding the local Frangi's index as voxel grey value. The local fibre orientation analysis results are represented by the 2nd order orientation tensor component a_{yy}. That is, in plane, orthogonal to the injection direction x. The tensor component is colour coded using a blue-to-red colour table with blue indicating values close to 0 and red close to 1. The misoriented central layer is clearly visible due to the high (red) values for a_{yy} there

3.2 Microstructure Generation

In the following the generation of virtual microstructures which have the same properties as the structures considered in Sect. 3.1.2 is described.

Stochastic volume elements with a fibre content of 20% are considered. The morphology of the glass fibres is described by the fibre length of 250 μm and the fibre diameter of 10 μm. The complete thickness of the 2 mm thick plate is resolved in order to capture the layered structure of the moulded specimens. The samples are described by 1024 voxels in each direction, i.e. the size of a voxel in each direction is obtained as 1.9313 μm. Thus each single fibre is resolved by five voxels over the thickness. The fibre orientation in each layer as well as the thickness of the misoriented layer are prescribed according to Table 3.1.

The realisation of sample 1 is depicted in Fig. 3.6. The size of the sample is 2 mm (1024 voxels) in each direction. Fibres belonging to the misoriented middle layer are

Table 3.1 Analysis results for the 3D images of samples 1–5, extracted at the positions shown in Fig. 3.1. The thickness of the central layers varies not only due to the varying strength of the misorientation but also due to the misoriented layers not being oriented perfectly parallel to the plate. Note that values of 1.00 or 0.00 for orientation vector components are due to rounding

Sample-layer	Volume (mm³)	Voxel size (μm)	# voxels	Orientation tensor diagonal elements			Main fibre orientation		
				a_{xx}	a_{yy}	a_{zz}	$\bar{v}_x$	$\bar{v}_y$	$\bar{v}_z$
1-1	1.82 × 2.21 × 0.99	1.3	1 400 × 1 700 × 760	0.68	0.27	0.04	−0.99	0.07	−0.00
1-2	1.88 × 2.47 × 0.09	1.3	1 450 × 1 900 × 67	0.22	0.73	0.06	0.17	−0.99	−0.01
1-3	1.82 × 2.34 × 0.84	1.3	1 400 × 1 800 × 650	0.66	0.28	0.06	1.00	−0.08	−0.01
2-1	1.56 × 0.86 × 0.90	1.2	1 300 × 720 × 750	0.69	0.26	0.05	1.00	0.10	−0.00
2-2	1.68 × 1.92 × 0.08	1.2	1 400 × 1 600 × 65	0.35	0.60	0.05	−0.47	0.88	−0.01
2-3	1.68 × 1.65 × 0.87	1.2	1 400 × 1 375 × 725	0.69	0.26	0.05	1.00	0.05	−0.00
3-1	2.47 × 2.08 × 0.78	1.3	1 900 × 1 600 × 600	0.60	0.35	0.04	−0.91	−0.41	−0.05
3-2	2.60 × 2.47 × 0.07	1.3	2 000 × 1 900 × 55	0.16	0.81	0.03	0.16	−0.99	−0.00
3-3	2.47 × 2.21 × 0.78	1.3	1 900 × 1 700 × 600	0.65	0.25	0.10	−0.99	−0.14	−0.03
4-1	1.56 × 1.62 × 0.90	1.2	1 300 × 1 350 × 750	0.66	0.28	0.05	−0.99	0.16	−0.00
4-2	1.56 × 1.62 × 0.06	1.2	1 300 × 1 350 × 51	0.40	0.58	0.02	−0.60	−0.81	−0.02
4-3	1.56 × 0.99 × 0.90	1.2	1 300 × 825 × 750	0.67	0.27	0.06	−0.98	0.18	0.00
5-1	1.80 × 1.46 × 0.90	1.2	1 500 × 1 215 × 750	0.68	0.26	0.06	−1.00	−0.02	0.00
5-2	1.80 × 1.98 × 0.11	1.2	1 500 × 1 650 × 93	0.20	0.77	0.03	0.16	−0.99	−0.00
5-3	1.77 × 1.14 × 0.90	1.2	1 475 × 950 × 750	0.69	0.24	0.06	1.00	−0.02	−0.01

highlighted in green. The fibres are distributed in such a way that the oriented and the misoriented layer are not separated in a strict manner. Fibres may overlap into the neighbouring layer.

In a next step the generated structures are compared to the corresponding samples of the μCT images from Sect. 3.1.1. Figures 3.7 and 3.8 display 2D slices for each sample from the upper and the middle layer. A comparison to the corresponding CT slices shows good agreement between the structures. It is noted that the virtual structures display a larger part of the microstructure (2 mm vs. 1.2 mm edge length) and therefore the fibres appear smaller than in the CT images.

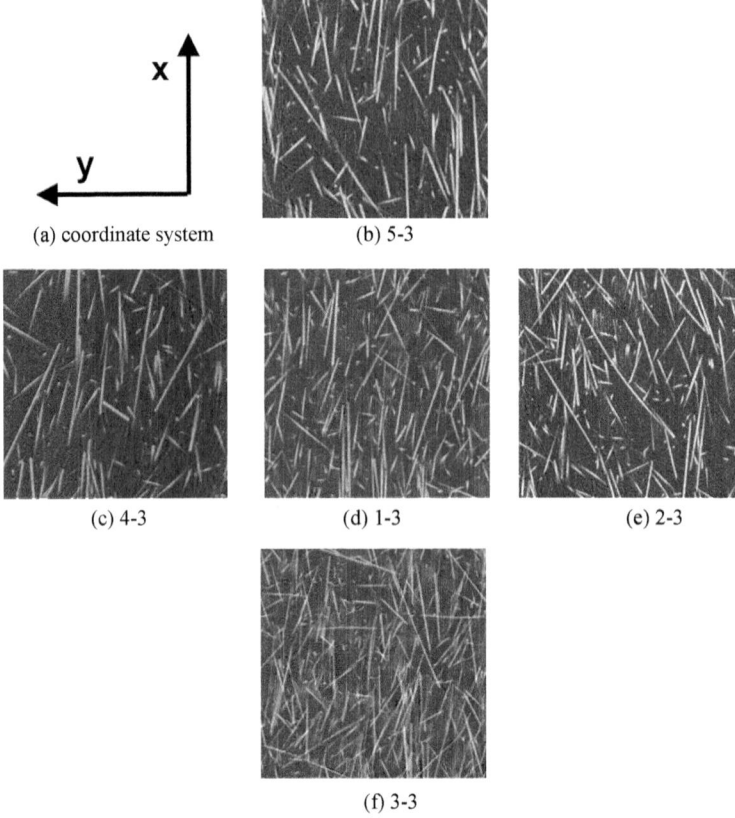

Fig. 3.4 1 000 × 1 000 pixel slices from the upper layers of samples 1–3 and 800 × 1 000 pixel slices from the upper layers of samples 4 and 5. Images are arranged as the sample extraction positions shown in Fig. 3.1

3.3 Identification of Material Parameters for the Matrix Material

Many methods for the characterisation of fibre reinforced composites are based on measurements of composite specimens, which have certain special fibre orientations, e.g. highly oriented or parallel fibres. In contrast, only measurements of the pure constituents are necessary in the presented method. This is an advantage especially in the case, if the constituents are isotropic or transversely anisotropic instead of fully anisotropic. In this section the method for the determination of material parameters and functions is described which are necessary to describe the rate-independent nonlinear material behaviour of the polymer matrix.

Polybutylene terephthalate (PBT), which is considered as matrix material in this book, is a thermoplastic semi-crystalline polymer and a type of polyester [1]. This

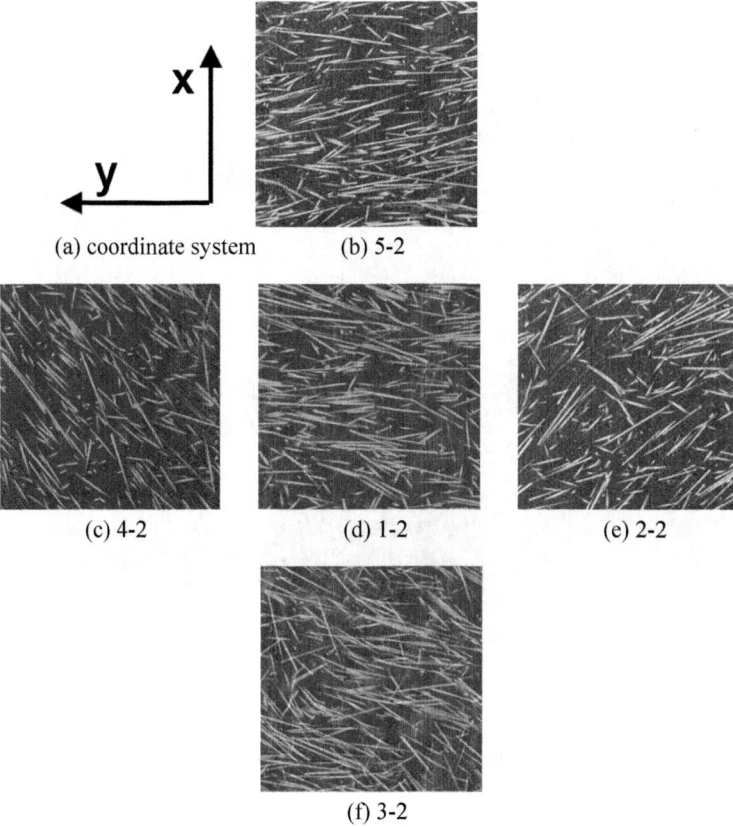

(a) coordinate system   (b) 5-2

(c) 4-2   (d) 1-2   (e) 2-2

(f) 3-2

Fig. 3.5 1 000 × 1 000 pixel slices from the misoriented central layers in samples 1–5. Images are arranged as the sample extraction positions shown in Fig. 3.1

polymer material shows a complex temperature-dependent viscoplastic behaviour with damage. However, this complex behaviour can be simplified to a time- and temperature-independent model for many applications. For the sake of simplicity, a standard time-independent elastoplasticity model with isotropic hardening and with a single additional internal variable

$$d = \frac{A_d}{A_0} \in [0, 1]$$

for isotropic damage is considered during the further procedure [6, 16, 19]. The damage variable d is defined as the share of the damaged surface area A_d on the total representative cross-section A_0. The damage variable d can be measured by several methods. The simplest method is to measure the degradation of the elastic modulus

Fig. 3.6 Visualisation with GeoDict [5] of sample 1, size of realisation 2 mm in each direction

$$d = 1 - \frac{E_d}{E_e}.$$

The degraded modulus can be determined in the unloading regime of cyclic tests (see Fig. 3.10).

The corresponding rate-independent material law with memory is introduced in Chap. 4

Remark 3.1 The viscoelastic damping, which is related to the area of the hysteresis loops is not taken into account and the different material parameters are determined for each testing speed. However, the model can be extended to capture viscoelastic or viscoplastic effects.

Now the elastoplastic material model with damage from Chap. 4 is repeated shortly. The free energy is decomposed into an elastic part $(1 - d) W_{\text{elastic}}$, a plastic part $(1 - d) W_{\text{plastic}}$, and a regularisation part W_{damage} resulting in

$$W(\varepsilon, d, \varepsilon_p, r) = (1 - d)(W_{\text{elastic}}(\varepsilon - \varepsilon_p) + W_{\text{plastic}}(\varepsilon_p, r)) + W_{\text{damage}}(d)$$

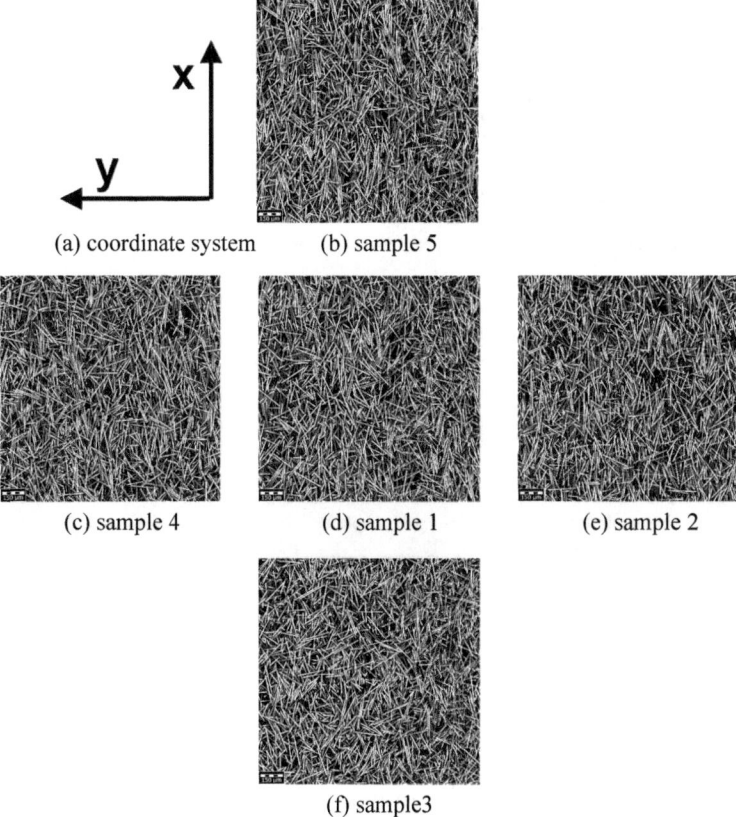

(a) coordinate system   (b) sample 5

(c) sample 4   (d) sample 1   (e) sample 2

(f) sample3

Fig. 3.7 $1\,024 \times 1\,024$ pixel ($2\,mm \times 2\,mm$) slices from the upper layers of virtual realisations of samples 1–5. Images are arranged as the sample extraction positions shown in Fig. 3.1

Both the elastic and plastic part are multiplied by the factor $(1 - d)$ for including the damage [6, 16]. The additional term W_{damage} guarantees $d < 1$. In the case of isotropic hardening the plastic part is written as a sum of both the linear hardening term and the Voce hardening term

$$W_{\text{plastic}}(\varepsilon_{\text{p}}, r) = W_{\text{iso}}(r) = \frac{1}{2}H_0 r^2 + (K_\infty - K_0)\left(r + \frac{e^{-\delta r}}{\delta}\right).$$

The plastic part W_{iso} depends on three positive material parameters

$$H_0 \geq 0, \qquad a_1 := K_\infty - K_0 \geq 0, \qquad \delta \geq 0 \tag{3.5}$$

which have to be fitted for the plastic behaviour. The expansion of the yield surface is then obtained as

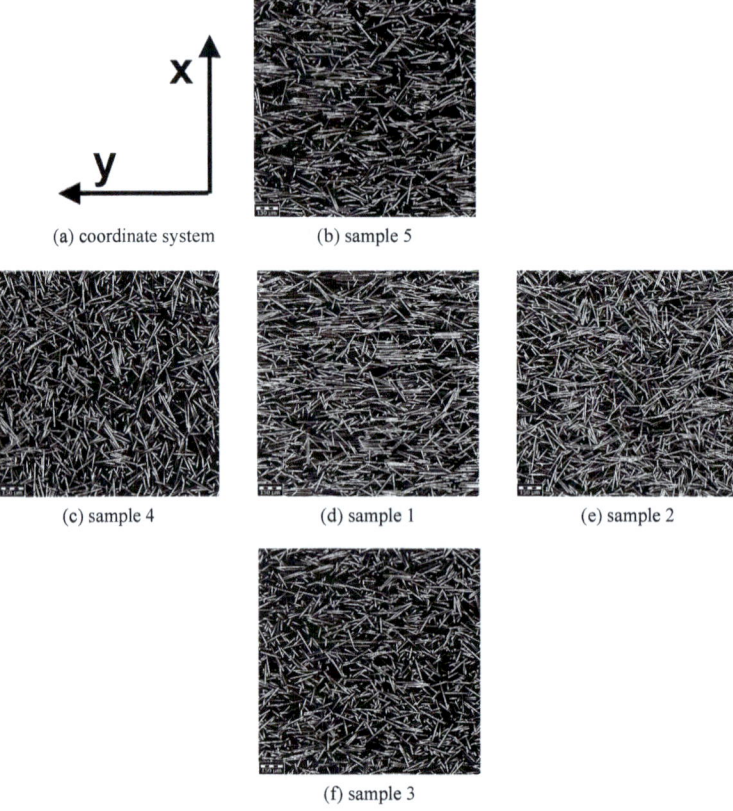

Fig. 3.8 1 024 × 1 024 pixel (2 mm × 2 mm) slices from the misoriented central layers of virtual realisations of samples 1–5. Images are arranged as the sample extraction positions shown in Fig. 3.1

Fig. 3.9 Measured
stress-strain curves for PBT
at testing speeds of 0.1, 1.0
and 10.0 mm/s

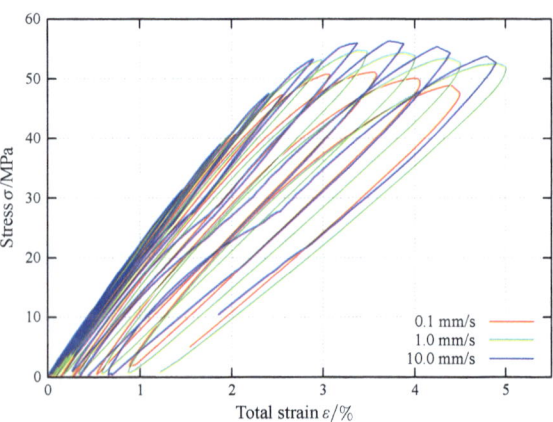

Fig. 3.10 Schematic
stress-strain diagram of the
first load cycle

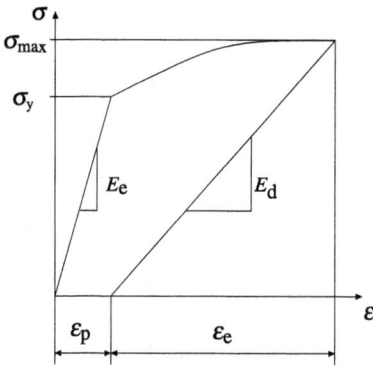

$$\psi(r) = \tilde{\psi}(r; \sigma_y, H_0, a_1, \delta) = \sigma_y + \frac{\partial W_{\text{iso}}}{\partial r} = \sigma_y + H_0 r + a_1 \left(1 - e^{-\delta r}\right),$$

where σ_y denotes the initial yield stress. The damage conjugated force is defined by

$$Y = -\frac{\partial W}{\partial d} = W_{\text{elastic}} + W_{\text{iso}}(r) + \frac{\partial W_{\text{damage}}}{\partial d}$$

$$= \frac{1}{2}\varepsilon_e : \mathbb{C} : \varepsilon_e + W_{\text{iso}}(r) + \frac{\partial W_{\text{damage}}}{\partial d}.$$

The last term can be neglected for small values of the damage variable d. Then the conjugated force Y is equal to the strain energy of the undamaged material. The damage accumulation is formulated as an explicit function

$$d = \tilde{d}(Y; Y_0, b1, b_2) = \frac{Y - Y_0}{Y} + b_1 \left(\frac{Y_0}{Y} - e^{-b_2(Y - Y_0)}\right) \tag{3.6}$$

of the damage conjugated force Y which contains three non-negative material parameters

$$Y_0 \geq 0, \qquad b_1 > 0, \qquad b_2 \geq 0. \tag{3.7}$$

The basis for the identification of the PBT material parameters are cyclic loading tests on pure PBT specimens which are explained in Chap. 6, see Fig. 3.9. Solely uniaxial tensile tests are considered in this section. Therefore, the isotropic hardening parameter r is identical with measured plastic strain ε_p, and the von Mises equivalent stress is given by $\sigma^{\text{eq}} \equiv \sigma$ in the spatial one-dimensional case, i.e. for the uniaxial tensile test. Each tensile test consists of several power-controlled loading-unloading cycles with increasing amplitudes. For cycle $k \in \{1, 2, \ldots, K\}$, the stress $\sigma^k = \sigma^k(t)$, $t \in (T^{k-1}, T^k]$ is prescribed and $\varepsilon^k = \varepsilon^k(t)$, $t \in (T^{k-1}, T^k]$ is measured (where $T^0 = 0$). Then (i) the maximum measured stress, (ii) the maximum measured total strain, (iii) the plastic strain, and (iv) the damage variable (see Fig. 3.10) are computed for each cycle $k \in \{1, 2, \ldots, K\}$:

$$\sigma_{\text{max}}^k = \max_{t \in (T^{k-1}, T^K]} \sigma^k(t), \qquad \varepsilon_{\text{max}}^k = \max_{t \in (T^{k-1}, T^K]} \varepsilon^k(t),$$

$$\varepsilon_p^k = \varepsilon^k(t = T^k), \qquad d^k = 1 - \frac{E_d^k}{E_e}.$$

The maximum elastic strain is taken as $\varepsilon_e^k := \varepsilon_{\text{max}}^k - \varepsilon_p^k$. The corresponding data $(\sigma_{\text{max}}^k, \varepsilon_e^k, \varepsilon_p^k, d^k)$ are listed in Table 3.2 for three testing speeds. The six unknown material parameters from (3.5), (3.7) and the unknown yield stress σ_y are computed by using the following fitting algorithm.

Algorithm 1: Parameter identification algorithm

Input : Young's modulus E_e, initial plastic strain ε_p^0

for $k \leftarrow 1$ **to** K **do**

  Compute effective stress $\bar{\sigma}^k = \frac{\sigma^k}{1-d^k}$

end

NLLS: fitting $\bar{\sigma} = \psi(r) = \tilde{\psi}(\varepsilon_p; \sigma_y, H_0, a_1, \delta)$ using data points $(\varepsilon_p^k, \bar{\sigma}^k)$

  Result: Plastic material parameter $\sigma_y, H_0, a_1, \delta$

for $k \leftarrow 1$ **to** K **do**

  Compute the damage conjugated force $Y^k = W_{\text{elastic}}(\varepsilon_e^k) + W_{\text{iso}}(\varepsilon_p^k) = \frac{1}{2} E_e(\varepsilon_e^k)^2 + \tilde{W}_{\text{iso}}(\varepsilon_p^k; \sigma_y, H_0, a_1, \delta)$

end

NLLS: fitting $d = \tilde{d}(Y; Y_0, b1, b_2)$ using data points (Y^k, d^k)

  Result: Damage material parameters Y_0, b_1, b_2

Table 3.2 Measured material parameters

Testing speed v/(mm/s)	Young's modulus E/MPa	Cycle number k	Maximum stress σ_{max}/MPa	Elastic strain ε_e/%	Plastic strain ε_p/%	Damage $d \in [0.1]$
0.1	2336	1	11.7	0.51	0.01	0.0
		2	22.4	1.02	0.03	0.03
		3	32.2	1.54	0.06	0.07
		4	40.7	2.05	0.09	0.11
		5	47.3	2.56	0.15	0.16
		6	50.7	3.07	0.28	0.22
		7	51.1	3.55	0.53	0.28
1.0	2385	1	11.2	0.49	0.02	0.0
		2	21.6	0.98	0.04	0.03
		3	31.4	1.48	0.05	0.08
		4	40.4	1.98	0.07	0.11
		5	47.8	2.48	0.11	0.16
		6	53.0	2.98	0.18	0.21
		7	54.5	3.38	0.34	0.25
10.0	2343	1	11.1	0.49	0.01	0.0
		2	21.7	0.99	0.05	0.01
		3	55.9	3.38	0.27	0.23
		4	56.2	3.74	0.46	0.27

3.3.1 Results

The determination of the Young's modulus and the Poisson's ratio for describing the the isotropic linear elastic behaviour of PBT is described in Chap. 6. The unknown parameters of the plastic yield $\psi = \tilde{\psi}(r; \sigma_y, H_0, a_1, \delta)$ function and the damage function $d = \tilde{d}(Y; Y_0, b1, b_2)$ are fitted by using a nonlinear least-squares (NLLS) Levenberg-Marquardt algorithm, see Algorithm 1. The initial plastic strain is zero at the testing speeds of 0.1 and 10.0 mm/s, whereas an initial plastic strain of 0.016% is estimated from the measurements at the testing speed of 1.0 mm/s. Furthermore, the parameter H_0 is set to zero for all testing speeds, because this parameter could not improve the fit of the hardening curve. The results of the parameter identification for three testing speeds are presented in Table 3.3. The good quality of the fits can be seen on Figs. 3.11, 3.12, 3.13, 3.14, 3.15 and 3.16. The Voce hardening parameter a_1

Table 3.3 Material parameter for the pure PBT polymer

Testing speed	v/(mm/s)	0.1	1.0	10.0
Elastic behaviour				
Young's modulus	E/MPa	2336	2385	2343
Poisson's ratio	v	0.4	0.4	0.4
von Mises J_2-plasticity with isotropic hardening				
Initial yield stress	σ_Y/MPa	3.32	3.24	1.00
Linear hardening	H_0/MPa	0.0	0.0	0.0
Voce hardening	a_1/MPa	66.31	69.34	79.34
	δ	1123.6	1592.4	785.7
Isotropic damage				
Threshold	Y_0/MPa	0.063	0.072	0.038
Parameter 1	b_1	1.007	0.965	1.010
Parameter 2	b_2/MPa^{-1}	0.295	0.225	0.255

Fig. 3.11 Stress and effective stress as function of the plastic strain at a testing speed of 0.1 mm/s (first fit)

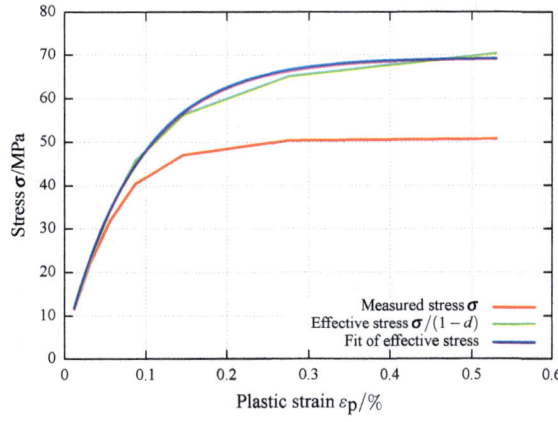

Fig. 3.12 Stress and effective stress as function of the plastic strain at a testing speed of 1.0 mm/s (first fit)

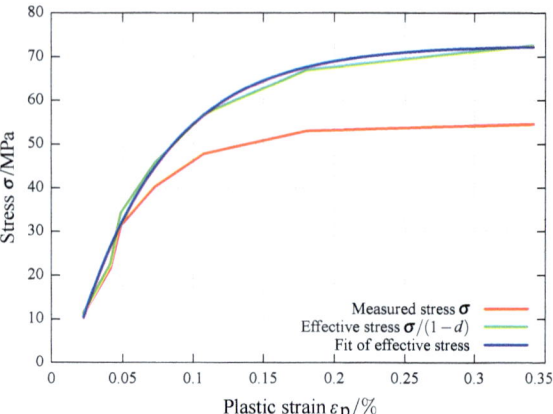

Fig. 3.13 Stress and effective stress as function of the plastic strain at a testing speed of 10.0 mm/s (first fit)

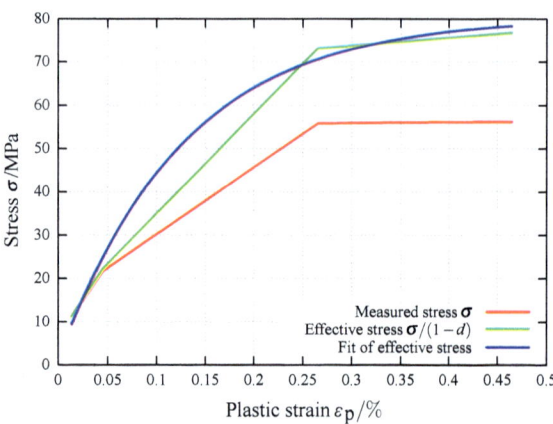

Fig. 3.14 Damage d as function of the energy release rate Y at a testing speed of 0.1 mm/s (first fit)

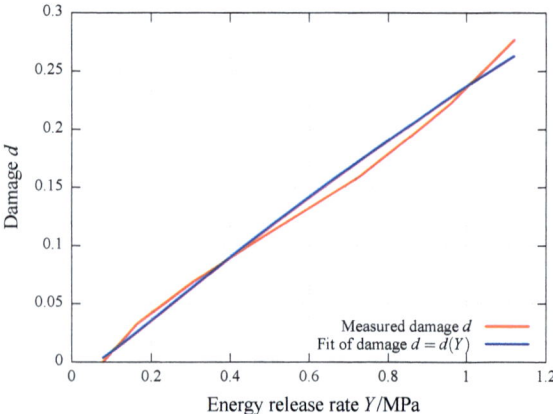

Fig. 3.15 Damage d as function of the energy release rate Y at a testing speed of 1.0 mm/s (first fit)

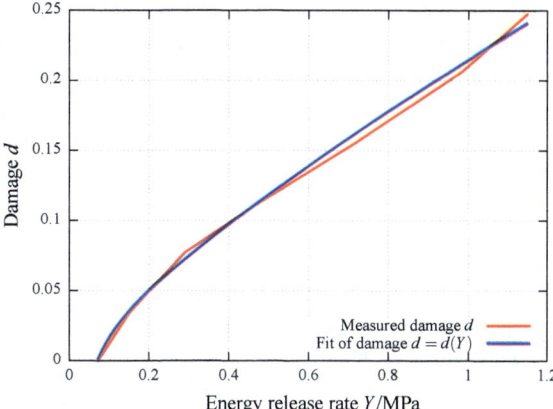

Fig. 3.16 Damage d as function of the energy release rate Y at a testing speed of 10.0 mm/s (first fit)

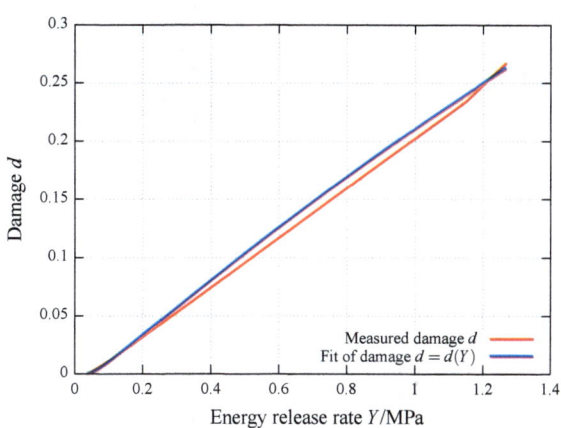

is increasing with the testing speed. The damage parameters are almost identical. The hardening curves are fitted very well also for very small plastic strains. However, the computed yield stress σ_y is also very small. Therefore, in the second fitting the initial yield stress σ_y is increased to 32.0 MPa, where the value is related to a plastic strain of about 0.05%, and only the remaining material parameters are fitted. At $\sigma = 32.0$ MPa the plastic strain is still below the value R_{p02}, which is usually used as initial yield stress for metallic materials. The initial plastic strain is set to zero for all testing speeds. The resulting elastoplastic material parameters of the second fit are presented in Table 3.4 and in Fig. 3.17. The Voce hardening parameter a_1 is increasing with the testing speed as in the first fit. The second fit of the plastic parameters can be used if small plastic strains are not of interest, see Fig. 3.17. Finally, the cyclic tensile tests are simulated by taking the identified material parameters from Table 3.3 (first fit) and Table 3.4 (second fit). The results of the first and second fit are visualised in Figs. 3.18 and 3.19 for the testing speed of 0.1 mm/s as well as in Figs. 3.20 and 3.21 for the testing speed of 10.0 mm/s. For both

Table 3.4 Material parameter for the pure PBT polymer

Testing speed	$v/(\text{mm/s})$	0.1	1.0	10.0
von Mises J_2-plasticity with isotropic hardening				
Initial yield stress	σ_Y/MPa	32.0	32.0	32.0
Linear Hardening	H_0/MPa	0.0	0.0	0.0
Voce hardening	a_1/MPa	39.96	43.56	45.45
	δ	642.6	819.0	876.12

Fig. 3.17 Stress and effective stress as function of the plastic strain at a testing speed of 0.1 mm/s (second fit)

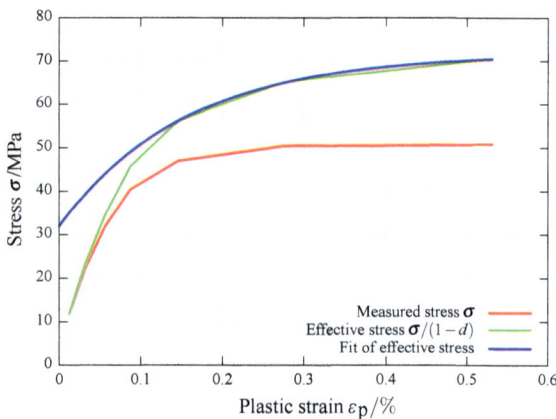

Fig. 3.18 Simulation and measurements of cyclic tensile tests, testing speed 0.1 mm/s (first fit)

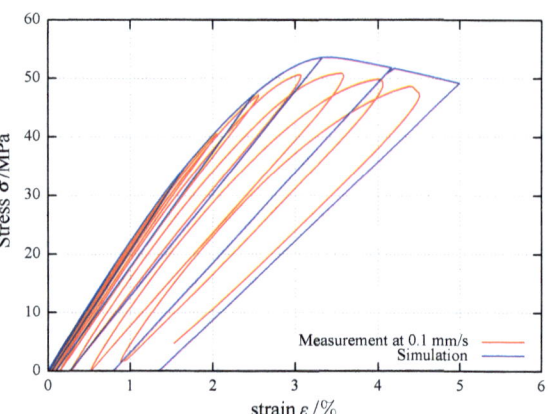

testing speeds, there is no essential difference between the simulation results using the plastic material parameters obtained from the first fit or second fit, respectively. A detailed sensitivity analysis for every material parameter is outside the scope of this section. The hysteresis loops (see Figs. 3.18, 3.19, 3.20 and 3.21) due to the viscoelastic behaviour of the PBT polymer are not captured by the time-independent elastoplastic material model with damage which is considered in this section.

Fig. 3.19 Simulation and
measurements of cyclic
tensile tests, testing speed
0.1 mm/s (second fit)

Fig. 3.19 Simulation and
measurements of cyclic
tensile tests, testing speed
0.1 mm/s (second fit)

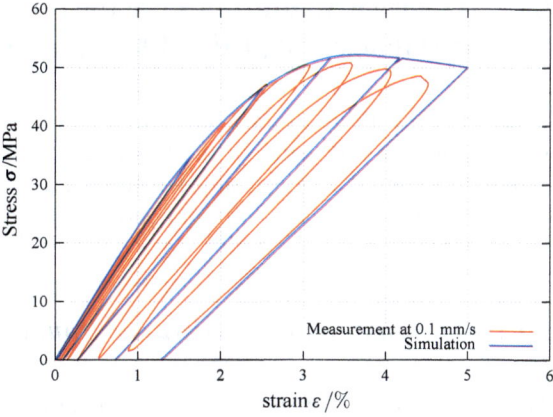

Fig. 3.20 Simulation and
measurements of cyclic
tensile tests, testing speed
10.0 mm/s (first fit)

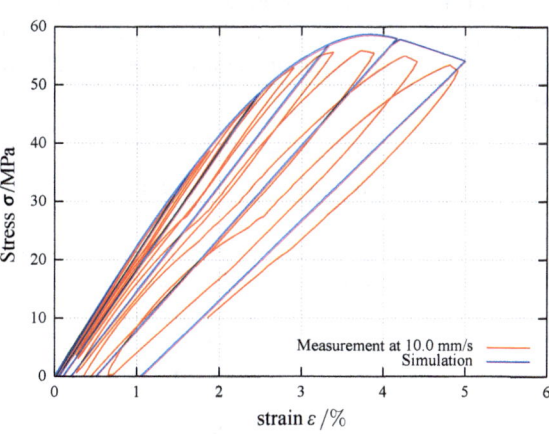

Fig. 3.21 Simulation and
measurements of cyclic
tensile tests, testing speed
10.0 mm/s (second fit)

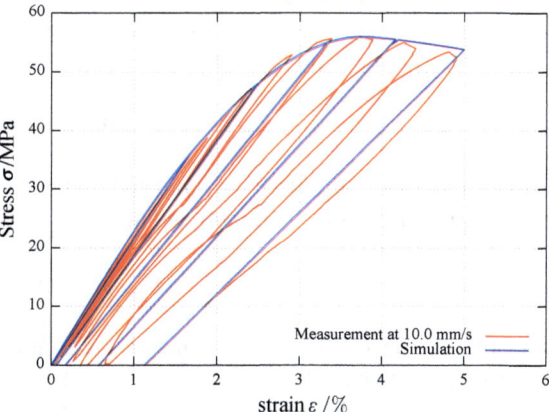

3.4 Numerical Solution of Periodic Boundary Problems for Damage Coupled to Elastoplasticity

In the following the governing equations of a periodic boundary value problem and their reformulation in terms of the Lippmann-Schwinger (LS) integral equations are briefly reviewed. Subsequently, the numerical solution of the periodic boundary value problem by application of fast Fourier transforms is outlined.

3.4.1 Formulation of the Periodic Boundary Value Problem

For the computation of the microscopic deformation a periodic boundary value problem (BVP) on a representative volume element (RVE) ω is considered. At the boundary $\partial\omega$ of the RVE an effective strain ε^M is applied. The kinematics for the unknown strains depending on the displacements $\mathbf{u}$ and the fluctuations $\mathbf{u}^*$ are given as

$$\left.\begin{aligned}\varepsilon(\mathbf{u})(\mathbf{x}) &= \varepsilon^M + \varepsilon(\mathbf{u}^*)(\mathbf{x})\\\varepsilon(\mathbf{u}^*)(\mathbf{x}) &= \tfrac{1}{2}\left(\text{grad } \mathbf{u}^*(\mathbf{x}) + \text{grad}^T \mathbf{u}^*(\mathbf{x})\right)\end{aligned}\right\}\mathbf{x} \in \omega. \tag{3.8}$$

At the boundary of the domain the following (anti-)periodic boundary conditions are prescribed in terms of the Cauchy stress σ as

$$\left.\begin{aligned}\mathbf{u}^*(\mathbf{x}) &\quad\# \\\sigma(\mathbf{x})\cdot\mathbf{n} &\quad -\#\end{aligned}\right\}\mathbf{x} \in \partial\omega. \tag{3.9}$$

Therein, # and −# denote periodicity or anti-periodicity, respectively. Therefore, the fluctuations at opposite faces of the RVE are equal, whereas the tractions have the same magnitude but point into opposite directions.

The equilibrium condition for the stresses read

$$\text{div } \sigma(\mathbf{x}) = \mathbf{0}, \ \mathbf{x} \in \omega. \tag{3.10}$$

The formulation of the BVP is completed by the constitutive functional $\mathscr{F}$, which connects the stresses to the strains and the history of the material via

$$\sigma(\mathbf{x}) = \mathscr{F}\left[\varepsilon(\mathbf{x}), \varepsilon_p(\mathbf{x}), r(\mathbf{x}), d(\mathbf{x})\right]. \tag{3.11}$$

According to Sect. 3.3, ε_p denotes the plastic strains and d the damage variable.

In a next step, the differential equation (3.10) for the stress equilibrium is reformulated into the so-called Lippmann-Schwinger integral equation, see [10], according to [23]. To this end the polarisation tensor τ is defined as

$$\tau(\mathbf{x}) = \sigma(\mathbf{x}) - \mathbb{C}^0 : \varepsilon(\mathbf{x}), \tag{3.12}$$

by introduction of a constant homogeneous reference stiffness tensor $\mathbb{C}^0$. The solution of the local equilibrium equation (3.10) can then be obtained by application the non-local Green's operator Γ^0 applied to the stress polarisation

$$\varepsilon(\mathbf{x}) = \varepsilon^M - \left(\Gamma^0 * \tau\right)(\mathbf{x}). \tag{3.13}$$

The Green's operator depends only on the homogeneous reference stiffness and the applied boundary conditions, and thus is independent of the strain fluctuations, see [9]. The convolution operator $*$ in equation (3.13) is defined as

$$\left(\Gamma^0 * \tau\right)(\mathbf{x}) = \int_\omega \Gamma^0(\mathbf{x} - \mathbf{y}) : \tau(\mathbf{y})\, d\mathbf{y}. \tag{3.14}$$

Finally, the nonlinear Lippmann-Schwinger integral equation is obtained as

$$\varepsilon^M = \varepsilon(\mathbf{x}) + \Gamma^0 * \left(\mathscr{F}\left[\varepsilon, \varepsilon_p, d\right] - \mathbb{C}^0 : \varepsilon\right)(\mathbf{x}). \tag{3.15}$$

Please note, that the Green's operator Γ^0 is independent of the fluctuations and therefore only depends on the linear elastic reference stiffness as well as on the boundary conditions, see [9]. In order to further simplify the notation in the following the LS equation is rewritten by application of the operator B_ε as

$$\varepsilon^M = \left((I + B_\varepsilon)\varepsilon\right)(\mathbf{x}). \tag{3.16}$$

Therein I denotes the identity operator δ_{ij}, where $\delta_{ij} = 1$ if $i = j$ and $\delta_{ij} = 0$ otherwise. The numerical solution of equation (3.16) is outlined in the following.

3.4.2 Numerical Solution of Lippmann-Schwinger Equation via Fast Fourier Transforms

The numerical solution of the LS integral equation as given in (3.16) can be obtained iteratively by using the Neumann series expansion for inverting the operator $I + B_\varepsilon(\mathbf{x})$. Thus, the iterates of the local strains are obtained as

$$\varepsilon^0(\mathbf{x}) = \varepsilon^M \tag{3.17}$$
$$\varepsilon^{k+1}(\mathbf{x}) = -B_\varepsilon(\mathbf{x})\varepsilon^k(\mathbf{x}) + \varepsilon^M, \quad k = 0, 1, 2, ... \tag{3.18}$$

These iterates can be computed efficiently by the so-called basis scheme as proposed by Moulinec and Suquet [11] for linear elastic material behaviour. An extension towards the account of nonlinear material behaviour is given by Moulinec and Suquet in [12].

The basis scheme consists of the following steps, which are repeated until convergence is reached:

1. Solve the constitutive equation in the real space and compute the stress polarisation

$$\tau^k = \sigma - \mathbb{C}^0 : \varepsilon.$$

2. Transformation of the stress polarisation into the Fourier space

$$\hat{\tau}^k = \text{FFT}\,(\tau)\,.$$

3. Update the strain field in the Fourier space by application of the Green's operator

$$\widehat{\varepsilon^{k+1}} = -\hat{\Gamma}^0 : \hat{\tau}^k.$$

4. Inverse Fourier transformation of the updated strain field

$$\varepsilon^{k+1} = \text{FFT}^{-1}\left(\widehat{\varepsilon^{k+1}}\right).$$

Explicit expressions for the Green's operator can be found e.g. in Mura [13]. Alternative solution schemes to the basis scheme which are also applicable to large deformations are summarised in Kabel et al. [7].

3.5 Computational Homogenisation

In the following the computational homogenisation scheme connecting the microscopic and the macroscopic scale is outlined. Here, focus is put onto classical first order homogenisation schemes, see e.g. [8] for details. Basically, the computational homogenisation scheme consists of the following four steps:

1. Generation of a representative volume element (RVE), see Sect. 3.2, and determination of the constitutive behaviour and material parameters of all phases.
2. Selection of admissible microscopic boundary conditions based on macroscopic input quantities.
3. Solution of microscopic boundary value problem (according to Sect. 3.4).
4. Determination of macroscopic output variables in terms of averaged microscopic quantities.

Here, the macroscopic variables are denoted by the index M. The effective macroscopic quantities, stresses or strains respectively, are obtained by averaging the corresponding microscopic solution fields over the volume $\|\omega\|$ of the RVE ω

$$\sigma^M = \frac{1}{|\omega|} \int_\omega \sigma \, dv \tag{3.19}$$

$$\varepsilon^M = \frac{1}{|\omega|} \int_\omega \varepsilon \, dv. \tag{3.20}$$

In the numerical examples given in the next section, so called mixed boundary conditions are prescribed to the RVE in order to reconstruct the performed experiments. In case of these boundary conditions macroscopic periodic strains ε^M are prescribed in loading direction (here 0° or 90°) and the other boundaries are stress-free. The effective stiffness in loading direction is then computed as the ratio of the macroscopic strain and stress in the corresponding loading direction.

3.6 Numerical Examples

In the following the results for the purely elastic and the elasto-plastic simulations are given and compared to the experimental data from Chap. 6.

In a first step the elastic behaviour is addressed. Therefore, the PBT matrix is modelled as an isotropic linear elastic material ($E = 2470$ MPa, $\nu = 0.4$). The PBT matrix is reinforced by 20% (weight) linear elastic glass fibres with a Young's modulus of 73,400 MPa and a Poisson's ratio 0.22. The simulations are carried out on the highly resolved (1024^3 voxels) virtual samples from Sect. 3.2. Mixed boundary conditions as described in Sect. 3.5 are applied.

In Fig. 3.22 the simulations and the experimental data are compared for the slowest measured loading. For the simulation of the 90° direction the simulation reflects the measurements well for all 5 regarded samples. In 0° direction the simulation underestimates the stiffness of the composite lightly for all regarded specimens.

Fig. 3.22 Comparison of simulation and measurement for elastic composite behaviour, Samples 1–5

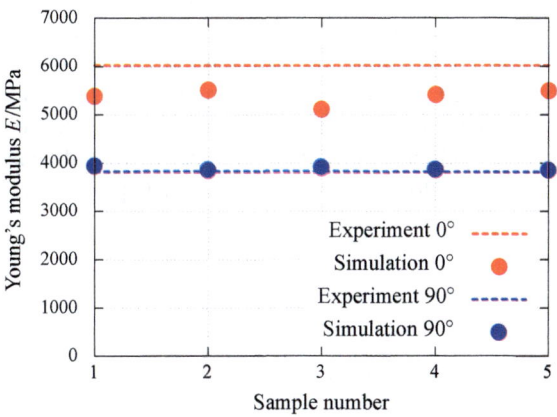

Fig. 3.23 RVE for
elasto-plastic simulation,
size $512 \times 512 \times 256$ voxel

Nevertheless, from Fig. 3.22 it is concluded that the simulations are in good coincidence with the experimental data. For all realisations similar results are obtained which is connected to the representativeness of the regarded volume elements.

A big advantage of the applied micromechanical approach is that not only the effective quantities like stresses σ^M and strains ε^M are available, but also the local fields in the RVE. In a next step the simulation of the elasto-plastic PBT model including damage is addressed. For this non-linear material behaviour a much higher resolution of the fibres is required. Therefore, only the upper layer of the multi-layered microstructure is considered, see Fig. 3.23. Herein, the fibres are resolved by 16 voxels over their thickness, i.e. the voxel size is chosen as 0.625 μm.

The effective elastic stiffnesses of this one-layered RVE result in $E_0^M = 5301$ MPa for the 0° direction and $E_{90}^M = 3920$ MPa for the 90° direction. Compared to the average stiffness of the layered RVE ($\bar{E}_0^M = 5412$ MPa and $\bar{E}_{90}^M = 3907$ MPa) and the experimental values ($E_0 = 6050$ MPa and $E_{90} = 3840$ MPa) very good results are obtained with the considered one-layered RVE.

For the elasto-plastic simulation the material parameters of the PBT matrix are chosen according to Table 3.4 for the plastic contribution and Table 3.3 for the damage. The simulation results for a loading and unloading scenario are depicted in Fig. 3.24. For the 90° direction very good coincidence of the simulation and the measurement are archived. For the 0° direction the simulation underestimates the stresses, which follows from the fact that for the considered RVEs also the elastic response yields lower stresses than the experiments.

Fig. 3.24 Comparison of simulation and experiment for elasto-plastic model including damage

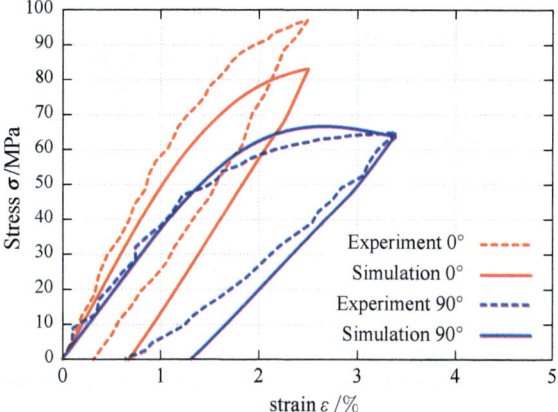

3.7   Conclusion

In summary, this section explains how the nonlinear behavior of the composite can be simulated in an RVE using only the material characterization of the pure polymer and the morphology of the microstructure.

Mechanical tests on the composite are not required.

References

1. Baur, E., Osswald, T.A., Rudolph, N., Brinkmann, S., Schmachtenberg, E. (eds.): Saechtling Kunststoff Taschenbuch, 31st edn. Hanser (2013)
2. Fisher, N., Lewis, T., Embleton, B.: Statistical Analysis of Spherical Data. Cambridge University Press, Cambridge, UK (1987)
3. Frangi, A., Niessen, W., Vincken, K., Viergever, M.: Multiscale vessel enhancement filtering. In: Proceedings of the Medical Image Computing and Computer-Assisted Intervention, pp. 130–137 (1998)
4. Fraunhofer ITWM, Department of Image Processing: MAVI—modular algorithms for volume images. http://www.mavi-3d.de (2005)
5. GeoDict.: www.geodict.com. Accessed 16 Jan 2019
6. Ju, J.: On energy-based coupled elastoplastic damage theories: constitutive modeling and computational aspects. Int. J. Solids Struct. **25**(7), 803–833 (1989)
7. Kabel, M., Böhlke, T., Schneider, M.: Efficient fixed point and Newton-Krylov solvers for FFT-based homogenization of elasticity at large deformations. Comput. Mech. **54**(6), 1497–1514 (2014)
8. Kouznetsova, V., Brekelmans, W., Baaijens, F.: An approach to micro-macro modeling of heterogeneous materials. Comput. Mech. **27**(1), 37–48 (2001)
9. Kröner, E.: Bounds for effective elastic moduli of disordered materials. J. Mech. Phys. Solids **25**(2), 137–155 (1977)
10. Lippmann, B., Schwinger, J.: Variational principles for scattering processes. Phys. Rev. **79**, 469–480 (1950)

11. Moulinec, H., Suquet, P.: A fast numerical method for computing the linear and nonlinear mechanical properties of composites. Comptes rendus de l'Académie des sciences. Série II, Mécanique, physique, chimie, astronomie **318**(11), 1417–1423 (1994)
12. Moulinec, H., Suquet, P.: A numerical method for computing the overall response of nonlinear composites with complex microstructure. Comput. Methods Appl. Mech. Eng. **157**(1–2), 69–94 (1998)
13. Mura, T.: Micromechanics of Defects in Solids, 2nd, revised edn. Mechanics of Elastic and Inelastic Solids. Martinus Nijhoff Publishers, Dordrecht (1987)
14. Niedziela, T., Strautins, U., Hosdez, V., Kech, A., Latz, A.: Improved multiscale fiber orientation modeling in injection molding of short fiber reinforced thermoplastics: simulation and Experiment. Int. J. Multiphys. Special Edition: Multiphys. Simul. Adv. Methods Ind. Eng. 357–366 (2011)
15. Ohser, J., Schladitz, K.: 3D Images of Materials Structures: Processing and Analysis. Wiley VCH (2009)
16. Onate, E. (ed.): Multiscale modeling of progressive damage in elasto-plastic composite materials (2014)
17. Otsu, N.: A threshold selection method from gray level histograms. IEEE Trans. Syst. Man Cybern. **9**, 62–66 (1979)
18. Schneider, R., Weil, W.: Stochastic and Integral Geometry. Probability and Its Applications. Springer, Heidelberg (2008)
19. Spahn, J., Andrä, H., Kabel, M., Müller, R.: A multiscale approach for modeling progressive damage of composite materials using fast Fourier transforms. Comput. Methods Appl. Mech. Eng. **268**, 871–883 (2014)
20. Stoyan, D., Kendall, W., Mecke, J.: Stochastic Geometry and Its Applications, 2nd edn. Wiley, Chichester (1995)
21. Wirjadi, O.: Models and algorithms for image-based analysis of microstructures. Ph.D. thesis. Technische Universität Kaiserslautern (2009)
22. Wirjadi, O., Schladitz, K., Easwaran, P., Ohser, J.: Estimating fibre direction distributions of reinforced composites from tomographic images. Image Anal. Stereol. **35**(3), 167–179 (2016)
23. Zeller, R., Dederichs, P.H.: Elastic constants of polycrystals. Phys. Status Solidi (b) **55**(2), 831–842 (1973)

Chapter 4
Parallel Inelastic Heterogeneous Multi-Scale Simulations

Ramin Shirazi Nejad and Christian Wieners

4.1 Introduction

Materials with micro-structures can be described effectively by homogenisation. Formally, the effective description can be obtained by a two-scale limit, where in every material point a periodic representative volume element (RVE) describing the material on the micro-scale is appended. Classical homogenisation computes an effective material response in one RVE and inserts the result into the macroscopic problem. This procedure is analytically well investigated, see, e.g., [2, 18, 22] for homogenisation and [1] for the two-scale limit. This is extended to heterogeneous materials, and for a corresponding numerical method the finite element convergence can be estimated with separate terms for the modelling and the approximation error on micro- and macro-scale [30].

For inelastic materials the heterogeneous multi-scale method introduced in [5, 13, 14, 26] is called the FE^2 method, see [23] for an overview. It is now well established for two-scale models in plasticity, damage and fracture. In case of rate-independent energetic material models [16] a full two-scale analysis for generalised standard materials with periodic coefficients is established in [17].

Here we apply this method to a short fibre reinforced material with a polymer matrix and glass fibre inclusions. Experimentally this material is investigated in [21], and appropriate models combining damage and plasticity effects are derived in [27, 28]. Here, the micro-structure requires a very fine resolution, so that for a full 3D simulation, e.g., cyclic loading in a tensile test, massive parallel computing is required.

Our contribution is the formulation of these models as rate-independent energetic material systems and the development of a fully scalable parallel algorithm for the incremental problem. This is realised in the parallel finite element software M++

R. S. Nejad · C. Wieners (✉)
Karlsruhe Institute of Technology, 76021 Karlsruhe, Germany
e-mail: christian.wieners@kit.edu

© Springer-Verlag GmbH Germany, part of Springer Nature 2019
S. Diebels and S. Rjasanow (eds.), *Multi-scale Simulation of Composite Materials*,
Mathematical Engineering, https://doi.org/10.1007/978-3-662-57957-2_4

[31] using multigrid methods for the solution of the linearisations on both scales. It includes a highly scalable periodic parallel direct coarse grid solver [12] for the linear system on the coarse level.

This work is organised as follows. We start in Sect. 4.2 with the heterogeneous two-scale finite element method for linear elasticity by defining the corresponding two-scale energy minimisation for the displacements on the macro-level and the micro-fluctuations in all RVEs. It is shown that this is equivalent to the classical homogenisation procedure where the macro-solution can be computed in two stages. In the first step, in the RVEs a basis of periodic micro-fluctuations corresponding to the symmetric tensor basis is computed, which then defines the homogenised Hookian tensor describing the effective material behaviour. This is inserted into the macro-problem which directly determines the homogenisation limit by the solution of the coarse mesh problem. For this standard procedure we propose an efficient parallel scheme, which is then evaluated numerically in Sect. 4.3 for a tensile test with different micro-structures. For the extension to inelastic models we introduce in Sect. 4.4 the analytic framework for rate-independent systems based on an energy functional depending on displacement and internal memory variables, and on a dissipation functional depending on the rate of the internal variables. Energy and dissipation are defined for small strain damage and plasticity models, the incremental stress response for these models are defined and the consistent tangent operator is derived explicitly. This is extended to nonlinear two-scale algorithms in Sect. 4.5, and finally in Sect. 4.6 the nonlinear numerical methods are evaluated by various test scenarios.

All results in this chapter are part of the PhD thesis of the first author [24], where the numerical realisation and the parallelisation strategy is explained in detail.

4.2   Parallel Heterogeneous Two-Scale FEM for Linear Elasticity

We introduce the two-scale method for elastic solids in $\Omega \subset \mathbb{R}^D$ in the case of small deformations. The method aims to approximate the deformation with a coarse mesh size H in a heterogeneous medium with much smaller characteristic length scale δ.

Small strain elasticity In the continuous problem the displacement vector $\mathbf{u}$ is characterised by minimising the total energy

$$\mathscr{E}(\mathbf{u}) = \int_{\Omega} W\big(\mathbf{x}, \boldsymbol{\varepsilon}(\mathbf{u})\big)\, d\mathbf{x} - \langle \ell, \mathbf{u} \rangle$$

subject to boundary conditions $\mathbf{u} = \mathbf{u}_D$ on $\Gamma_D \subset \partial\Omega$. Here, a load functional

$$\langle \ell, \mathbf{v} \rangle = \int_{\Omega} \mathbf{b}(\mathbf{x}) \cdot \mathbf{v}(\mathbf{x})\, d\mathbf{x} + \int_{\Gamma_N} \mathbf{t}_N(\mathbf{x}) \cdot \mathbf{v}(\mathbf{x})\, d\mathbf{a}$$

with body forces $\mathbf{b}$ and surface tractions $\mathbf{t}_N$ is applied, and the small strain isotropic elastic energy is of the form

$$W(\mathbf{x}, \boldsymbol{\varepsilon}) = \frac{1}{2}\mathbb{C}(\mathbf{x})[\boldsymbol{\varepsilon}(\mathbf{x})] : \boldsymbol{\varepsilon}(\mathbf{x}) \qquad (4.1)$$

only depending on the linearised strain $\boldsymbol{\varepsilon} = \boldsymbol{\varepsilon}(\mathbf{u})$ with $\boldsymbol{\varepsilon}(\mathbf{u}) = \text{sym } \mathbf{Du}$.

We consider the case that $\mathbb{C}(\cdot)$ is strongly inhomogeneous and cannot be resolved on coarse meshes of mesh size H, i.e., we consider the case that $\mathbb{C}(\cdot)$ can be resolved only on a mesh size $h < \delta$ with $\delta \ll H$, so that it is not feasible to compute the full fine mesh solution in Ω.

The multi-scale idea A multi-scale method aims to approximate the exact solution $\mathbf{u}$ on the coarse level in a finite element space with mesh size parameter $H > 0$, where the coarse approximation $\mathbf{u}_H$ (referred as macro-solution) is obtained by solving a suitable averaged problem. Therefore, we define a suitable averaged energy $\mathscr{E}_H(\cdot)$ so that the coarse approximation can be determined as minimiser of this energy.

The construction of the averaged energy relies on the solution of local problems in *representative volume elements* (RVE)

$$\mathscr{Y}_\xi = \xi + \delta(-0.5, 0.5)^D \subset \Omega$$

at sample points $\xi \in \varXi_H \subset \Omega$. On the RVEs we define locally micro-solutions

$$\mathbf{u}_{\xi,h} = \mathbf{u}_{\xi,H} + \mathbf{v}_{\xi,h},$$

where $\mathbf{u}_{\xi,H}$ is the linearisation of the macro-solution and $\mathbf{v}_{\xi,h}$ is the so-called micro-fluctuation. This is approximated in a finite element space with mesh size parameter $h > 0$. The two-scale setting is illustrated in Fig. 4.1.

The heterogeneous multi-scale method extends the two-scale homogenisation of periodic micro-structures to applications, where the micro-structure in every RVE is representative at least for a small neighbourhood extending the RVE. It is a modelling assumption that a periodic continuation of the micro-fluctuation is appropriate. These assumptions are quite restrictive but they allow for a full mathematical analysis of the homogenisation error which is enhanced by the modelling error due to the approximation of the heterogeneous micro-structure [30].

The discrete multi-scale setting Let $\Omega \subset \mathbb{R}^D$ be a Lipschitz domain, and define the space $V = H^1(\Omega; \mathbb{R}^D)$. For given Dirichlet data $\mathbf{u}_D$, we define the affine space $V(\mathbf{u}_D) = \{\mathbf{v} \in V : \mathbf{v} = \mathbf{u}_D \text{ on } \varGamma_D \subset \partial\Omega\}$. The macro-solution is approximated in a finite element space $V_H \subset V$, and we set

$$V_H(\mathbf{u}_D) = \{\mathbf{v}_H \in V_H : \mathbf{v}_H(\mathbf{x}) = \mathbf{u}_D(\mathbf{x}) \text{ for all nodal points } \mathbf{x} \in \varGamma_D\}.$$

On the mesh corresponding to V_H we select quadrature points $\varXi_H \subset \Omega$ with weights ω_ξ for $\xi \in \varXi_H$, and we introduce the notation

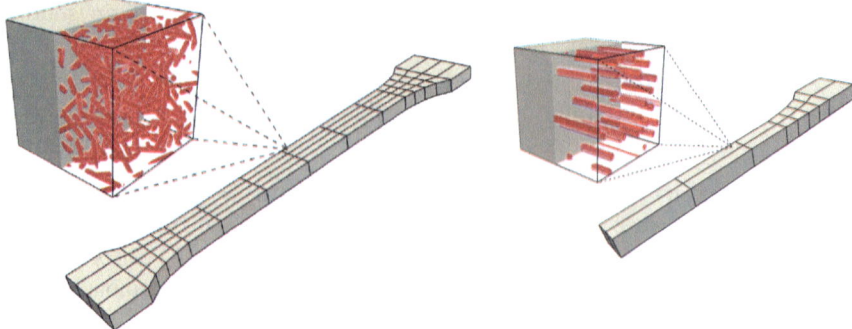

Fig. 4.1 Illustrations of the two-scale model with isotropic and unidirectional fibre directions: In the domain Ω, the macroscopic solution $\mathbf{u}_H$ is approximated on a coarse scale with mesh size H. The micro-structure of size $\delta \ll H$ is approximated at sample points ξ. The micro-fluctuation $\mathbf{v}_{\xi,h}$ and the effective material response is computed in the representative volume element $\mathscr{Y}_\xi = \xi + \delta(-0.5, 0.5)^D \subset \Omega$. The symmetries of the tensile test can be exploited to reduce the computational domain for the approximation of the macro-solution

$$\int_{\Xi_H} f(\xi) = \sum_{\xi \in \Xi_H} \omega_\xi f(\xi) \approx \int_\Omega f(\mathbf{x}) \, \mathrm{d}\mathbf{x}.$$

We assume that the quadrature is exact for $\boldsymbol{\varepsilon}(\mathbf{u}_H)$.

Locally in every RVE, the micro-fluctuation is approximated in a finite element space $V_{\xi,h} \subset V_\xi$ with

$$V_\xi = \left\{ \mathbf{v}_\xi \in \mathrm{H}^1_{\mathrm{per}}(\mathscr{Y}_\xi, \mathbb{R}^D) : \int_{\mathscr{Y}_\xi} \mathbf{v}_\xi \, \mathrm{d}\mathbf{x} = 0 \right\},$$

where $\mathrm{H}^1_{\mathrm{per}}(\mathscr{Y}_\xi, \mathbb{R}^D)$ denotes the restriction of $\mathscr{Y}_\xi$-periodic $\mathrm{H}^1_{\mathrm{loc}}(\mathbb{R}^D, \mathbb{R}^D)$ functions to $\mathscr{Y}_\xi$. On the micro-scale we set the global finite element approximation space as the product space $V_h = \prod_{\xi \in \Xi_H} V_{\xi,h}$. Furthermore, we assume that the elasticity tensor $\mathbb{C}(\mathbf{x})$ for $\mathbf{x} \in \mathscr{Y}_\xi$ is representative for the material properties in a neighbourhood of any sample point $\xi \in \Xi_H$.

The multi-scale problem The multi-scale approximation represented by the macro-solution and the micro-fluctuations in every RVE is defined as the minimiser $(\mathbf{u}_H, \mathbf{v}_h) \in V_H(\mathbf{u}_\mathrm{D}) \times V_h$ of the two-scale energy

$$\mathscr{E}_H(\mathbf{u}_H, \mathbf{v}_h) = \int_{\Xi_H} W_\xi(\boldsymbol{\varepsilon}(\mathbf{u}_H), \mathbf{v}_{\xi,h}) - \langle \ell, \mathbf{u}_H \rangle,$$

where the micro-energy is evaluated on the RVEs by

$$W_\xi(\boldsymbol{\varepsilon}_H, \mathbf{v}_{\xi,h}) = \frac{1}{|\mathscr{Y}_\xi|} \int_{\mathscr{Y}_\xi} W\big(\mathbf{x}, \boldsymbol{\varepsilon}_H(\xi) + \boldsymbol{\varepsilon}(\mathbf{v}_{\xi,h})\big) \, \mathrm{d}\mathbf{x}$$

depending on the strain $\varepsilon_H = \varepsilon(\mathbf{u}_H)$ of the macro-solution and micro-fluctuations $\mathbf{v}_{\xi,h}$. In the RVE we define the linear approximation of the macro-solution by $\mathbf{u}_{\xi,H}(\mathbf{x}) = \mathbf{u}_H(\xi) + \mathrm{D}\mathbf{u}_H(\xi)(\mathbf{x} - \xi)$. This defines, together with the micro-fluctuation, the micro-solution $\mathbf{u}_{\xi,h} = \mathbf{u}_{\xi,H} + \mathbf{v}_{\xi,h}$, so that by construction $\mathbf{u}_{\xi,h} - \mathbf{u}_{\xi,H}$ is periodic, and the strain of the macro-solution $\varepsilon_{\xi,H} = \varepsilon(\mathbf{u}_{\xi,H}) \equiv \varepsilon(\mathbf{u}_H)(\xi)$ is constant in the RVE $\mathscr{Y}_\xi$.

The two-scale problem and the multi-scale tensor The minimiser of the two-scale energy is characterised as the critical point of the two-scale energy: find $(\mathbf{u}_H, \mathbf{v}_h) \in V_H(\mathbf{u}_D) \times V_h$ satisfying

$$\textbf{Macro-Equilibrium} \quad 0 = \partial_u \mathscr{E}_H(\mathbf{u}_H, \mathbf{v}_h), \quad (4.2a)$$

$$\textbf{Micro-Equilibrium} \quad 0 = \partial_v \mathscr{E}_H(\mathbf{u}_H, \mathbf{v}_h), \quad (4.2b)$$

i.e., solving the coupled linear problems

$$\sum_\xi \frac{\omega_\xi}{|\mathscr{Y}_\xi|} \int_{\mathscr{Y}_\xi} \mathbb{C}(\mathbf{x})[\varepsilon(\mathbf{u}_H)(\xi) + \varepsilon(\mathbf{v}_{\xi,h})(\mathbf{x})] : \varepsilon(\delta\mathbf{u}_H)(\xi)\, \mathrm{d}\mathbf{x} = \langle \ell, \delta\mathbf{u}_H \rangle, \quad (4.3a)$$

$$\frac{1}{|\mathscr{Y}_\xi|} \int_{\mathscr{Y}_\xi} \mathbb{C}(\mathbf{x})[\varepsilon(\mathbf{u}_H)(\xi) + \varepsilon(\mathbf{v}_{\xi,h})(\mathbf{x})] : \varepsilon(\delta\mathbf{v}_{\xi,h})\, \mathrm{d}\mathbf{x} = 0 \quad (4.3b)$$

for test functions $(\delta\mathbf{u}_H, \delta\mathbf{v}_{\xi,h}) \in V_H(\mathbf{0}) \times V_{\xi,h}$.

Now we reduce this system to an averaged macro-problem. Therefore, we introduce an orthonormal basis $\eta_1, \ldots \eta_6$ of $\mathrm{Sym}(3) = \mathbb{R}^{3\times3}_{\mathrm{sym}}$. Corresponding to this basis we compute micro-fluctuations $\mathbf{w}_{\xi,h,1}, \ldots, \mathbf{w}_{\xi,h,6} \in V_{\xi,h}$ solving

$$\frac{1}{|\mathscr{Y}_\xi|} \int_{\mathscr{Y}_\xi} \mathbb{C}(\mathbf{x})[\eta_j + \varepsilon(\mathbf{w}_{\xi,h,j})(\mathbf{x})] : \varepsilon(\delta\mathbf{v}_{\xi,h})(\mathbf{x})\, \mathrm{d}\mathbf{x} = 0, \quad \delta\mathbf{v}_{\xi,h} \in V_{\xi,h}. \quad (4.4)$$

Inserting the representation of the macro-strain $\varepsilon_{\xi,H} = \varepsilon(\mathbf{u}_H)(\xi)$ with respect to this basis

$$\varepsilon_{\xi,H} = \sum_{j=1}^{6} \left(\varepsilon_{\xi,H} : \eta_j \right) \eta_j,$$

we obtain for the micro-fluctuation solving (4.3b)

$$\mathbf{v}_{\xi,h} = \sum_{j=1}^{6} \left(\varepsilon_{\xi,H} : \eta_j \right) \mathbf{w}_{\xi,h,j}.$$

This is now inserted into the macro-equation (4.3a), which yields

$$\int_{\varXi_H} \mathbb{C}_{\xi,H}[\boldsymbol{\varepsilon}(\mathbf{u}_H)] : \boldsymbol{\varepsilon}(\delta\mathbf{u}_H) = \langle \ell, \delta\mathbf{u}_H \rangle, \qquad \delta\mathbf{u}_H \in V_H(\mathbf{0})$$

with the two-scale elasticity tensor

$$\mathbb{C}_{\xi,H} = \sum_{j,k=1}^{6} \left(\frac{1}{|\mathscr{Y}_\xi|} \int_{\mathscr{Y}_\xi} \mathbb{C}(\mathbf{x})[\eta_j + \boldsymbol{\varepsilon}(\mathbf{w}_{\xi,h,j})(\mathbf{x})] : \eta_k \, d\mathbf{x} \right) \eta_j \otimes \eta_k . \tag{4.5}$$

Together, we obtain the following result.

Lemma 4.1 *The macro-solution* $\mathbf{u}_H \in V_H(\mathbf{u}_D)$ *of the heterogeneous multi-scale method minimises the averaged energy*

$$\mathscr{E}_H^{avg}(\mathbf{u}_H) = \frac{1}{2} \int_{\varXi_H} \mathbb{C}_{\xi,H}[\boldsymbol{\varepsilon}(\mathbf{u}_H)] : \boldsymbol{\varepsilon}(\mathbf{u}_H) - \langle \ell, \mathbf{u}_H \rangle . \tag{4.6}$$

The definition of the micro-fluctuations $\mathbf{w}_{\xi,h,1}, \ldots \mathbf{w}_{\xi,h,6}$ by Eq. (4.4) ensures that the two-scale elasticity tensor $\mathbb{C}_{\xi,H}$ is symmetric. Due to Korn's inequality the averaged energy is uniformly convex, which assures the existence and uniqueness of the minimiser in (4.6).

The parallel two-scale model In our parallel model we assume that the meshes to resolve the geometry in the RVEs are so large that they can be distributed to all processes. On the other hand, we do expect that we do not need to compute a micro-fluctuation in all RVEs $\mathscr{Y}_\xi$, e.g., if the micro-structure is identical. In the most simple case of two-scale homogenisation, we compute the micro-problem only once, as it is now described for our first parallel two-scale FEM of the heterogeneous small strain elasticity problem.

For the implementation, the RVEs $\mathscr{Y}_\xi = \xi + \delta(-0.5, 0.5)^D \subset \varOmega$ are mapped to the unit cube $\mathscr{Y} = (0, 1)^3$. In this simple case the micro-structure described by the elasticity tensor $\mathbb{C}$ is mapped to the same tensor for all $\xi \in \varXi_H$. Nevertheless, since this cannot be expected for general applications, we describe the parallel algorithm in a more general case which allows for different micro-structures and which also is flexible for cases where only some of the RVE computations are required.

Here we consider the case, that the approximation of the micro-structure requires a very fine mesh size, so that also the micro-problem has to be computed in parallel. Therefore, we determine by a load balancing procedure a domain decomposition $\bar{\varOmega} = \bar{\varOmega}^1 \cup \cdots \cup \bar{\varOmega}^P$ and a further decomposition for the reference RVE $\bar{\mathscr{Y}} = \bar{\mathscr{Y}}^1 \cup \cdots \cup \bar{\mathscr{Y}}^P$. The finite element spaces V_H and $V_{\xi,h}$ are distributed to the processes $p \in \mathscr{P} = \{1, \ldots, P\}$ which results into a consistent representation of the macro-deformation $\mathbf{u}_H$ and the micro-fluctuations $\mathbf{w}_{\xi,h,1}, \ldots, \mathbf{w}_{\xi,h,6}$ by local functions

$$\mathbf{u}_H^p = \mathbf{u}_H|_{\bar{\varOmega}^p}$$

and

$$\mathbf{w}_{\xi,h,j}^p = \mathbf{w}_{\xi,h,j}|_{\bar{\mathscr{Y}}^p}$$

on process p. For the elastic two-scale solution the multi-scale tensor $\mathbb{C}_{\xi,H}$ is evaluated only for a subset $\varXi_H^{\text{active}}$ with different micro-structures, i.e., we assume that for all other points $\xi \in \varXi_H \setminus \varXi_H^{\text{active}}$ some active point $\xi' \in \varXi_H^{\text{active}}$ exists so that we can choose $\mathbb{C}_{\xi,H} = \mathbb{C}_{\xi',H}$. The full parallel two-scale method is summarised in Algorithm 2.

Algorithm 2: Parallel heterogeneous two-scale method for linear elasticity.

E1) Sequentially for all points $\xi \in \varXi_H^{\text{active}}$ with different micro-structure perform the following steps:

    M1) Compute the micro-fluctuations $\mathbf{w}_{\xi,h,1}, \ldots, \mathbf{w}_{\xi,h,6} \in V_{\xi,h}$ solving in parallel

$$\frac{1}{|\mathscr{Y}_{\xi}|} \int_{\mathscr{Y}_{\xi}} \mathbb{C}(\mathbf{x})[\eta_l + \boldsymbol{\varepsilon}(\mathbf{w}_{\xi,h,l})] : \boldsymbol{\varepsilon}(\delta\mathbf{v}_{\xi,h}) \, d\mathbf{x} = 0, \qquad \delta\mathbf{v}_{\xi,h} \in V_{\xi,h}.$$

    M2) Evaluate the local contributions of the multi-scale tensor

$$\mathbb{C}_{\xi,H}^p = \frac{1}{|\mathscr{Y}_{\xi}|} \sum_{l,j=1}^{6} \left(\int_{\mathscr{Y}_{\xi}^p} \mathbb{C}(\mathbf{x})[\eta_l + \boldsymbol{\varepsilon}(\mathbf{w}_{\xi,h,l}^p)] : \eta_j \, d\mathbf{x} \right) \eta_l \otimes \eta_j, \quad p \in \mathscr{P}.$$

    M3) On the process q with $\xi \in \varXi_H \cap \Omega^q$ collect the full multi-scale tensor

$$\mathbb{C}_{\xi,H} = \sum_{p=1}^{P} \mathbb{C}_{\xi,H}^p.$$

E2) Sequentially for all points $\xi \in \varXi_H \setminus \varXi_H^{\text{active}} \cap \Omega^p$ find $\xi' \in \varXi_H^{\text{active}}$ with $\mathbb{C}_{\xi,H} = \mathbb{C}_{\xi',H}$ and send the multi-scale tensor to process p.

E3) Compute $\mathbf{u}_H \in V_H(\mathbf{u}_{\mathrm{D}})$ solving in parallel

$$\int_{\varXi_H} \mathbb{C}_{\xi,H}[\boldsymbol{\varepsilon}(\mathbf{u}_H)] : \boldsymbol{\varepsilon}(\delta\mathbf{u}_H) = \langle \ell, \delta\mathbf{u}_H \rangle, \qquad \delta\mathbf{u}_H \in V_H(\mathbf{0}).$$

4.3 Numerical Experiments for Linear Elastic Two-Scale Models

We evaluate the two-scale method for two component composites with the thermoplastic polymer *polybutylene terephthalate* as carrier matrix and embedded glass fibres. We use isotropic linear elasticity with Lamé parameters $\lambda_{\mathrm{M}} = 3571.43$ and $\mu_{\mathrm{M}} = 892.857$ for the polymer,[1] and $\lambda_{\mathrm{F}} = 30\,000$ and $\mu_{\mathrm{F}} = 20\,000$ for the glass fibre.[2]

For the test scenario we use a standardised uniaxial tensile test configuration ISO 527-2:1996 type 1A, where experimental data are available [4, 21]. The com-

---

[1] BASF data sheet on http://www.plasticsportal.net.
[2] Data sheet on http://www.matweb.com.

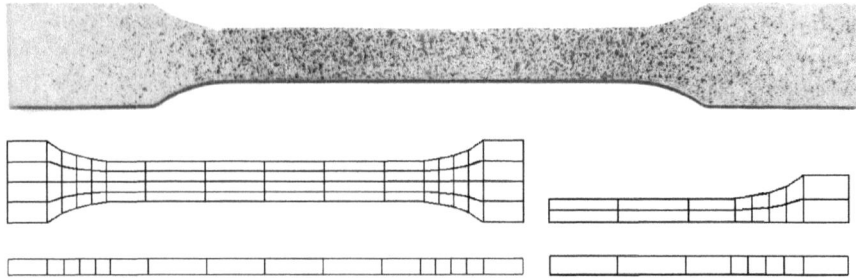

Fig. 4.2 Approximation on the coarse level of the geometry of the standardised specimen with 20 mm shortened shoulders and reduction to one-fourth

putational domain $\Omega \subset (-0.2, 0.2) \times (-2, 2) \times (-6.5, 6.5)$ is approximated by a hexahedral mesh, cf. Fig. 4.2.

Corresponding to the experimental setting we prescribe the displacement at the Dirichlet boundary $\Gamma_D = \{\mathbf{x} \in \partial\Omega : x_3 = -6.5 \text{ or } x_3 = 6.5\}$ with

$$\mathbf{u}_D(t, \mathbf{x}) = \begin{pmatrix} 0 \\ 0 \\ u_0 t \end{pmatrix} \quad \text{for } x_3 = 6.5, \qquad \mathbf{u}_D(t, \mathbf{x}) = 0 \text{ for } x_3 = -6.5,$$

$$\sigma(\varepsilon(\mathbf{u}))\mathbf{n} = 0 \quad \text{on } \Gamma_N = \partial\Omega \setminus \Gamma_D. \tag{4.7}$$

The scaling factor is set to $u_0 = 0.01$, and the linear model is tested for $t = 1$. For the investigation of the convergence properties, we reduce the computation to one fourth of the geometry $\Omega_{\text{sym}} \subset \Omega$ with symmetry boundary conditions on $\Gamma_{\text{sym}} = \{\mathbf{x} \in \partial\Omega_{\text{sym}} : x_3 = 0 \text{ or } x_2 = 0\}$ which corresponds to two-sided loading with u_0 replaced by $0.5u_0$.

In V_H and in $V_{\xi,h}$ we use conforming hexahedral $\mathbb{Q}_1$ finite elements.

Numerical tests for different resolutions of the micro-structure In our first experiment, the convergence of the classical two-scale homogenisation is tested, where only one representative micro-structure is computed, and where the same effective material response is used at all integration points. Therefore, we select from the collection [8] one micro-structure $\mathscr{Y}_\xi^\delta$ of characteristic length scale δ with isotropic glass fibre distribution and 10% fibre volume fraction. Corresponding to a basis of the symmetric tensors, 6 periodic micro-fluctuations are computed by (4.4), see Fig. 4.3.

This defines by (4.5) the effective elasticity tensor

$$\mathbb{C}_{\xi,H}^\delta = \begin{pmatrix} 6813.81 & 4103.21 & 4091.94 & -31.99 & -1.79 & -30.05 \\ 4103.21 & 6746.61 & 4090.63 & -1.79 & 26.43 & -3.04 \\ 4091.94 & 4090.63 & 6750.23 & 9.40 & 2.73 & -22.82 \\ -31.99 & -1.79 & 9.40 & 2630.59 & -7.91 & -1.96 \\ -1.79 & 26.43 & 2.73 & -7.91 & 2598.53 & 10.05 \\ -30.05 & -3.04 & 22.82 & -1.96 & 10.05 & 2597.09 \end{pmatrix}.$$

Fig. 4.3 Deformation of the 6 periodic micro-fluctuations $\mathbf{w}_{\xi,h,k}^{\delta}$ and the Frobenius norm distribution of the stress $|\sigma_{\xi,h,k}^{\delta}|$ in $\mathscr{Y}_{\xi}^{\delta}$ corresponding to the symmetric tensor basis $\eta_1, \ldots, \eta_6$

The accuracy of the two-scale approximation depends on the mesh size of the macro- and the micro-problem, and on the resolution δ of the geometry in the RVE. Since the complex micro-structure in the RVE requires a very fine resolution of the micro-problem, we test if smaller RVEs $\mathscr{Y}_{\xi}^{\delta/2}$ and $\mathscr{Y}_{\xi}^{\delta/4}$ with a coarser resolution of the micro-problem are sufficient.

For the quantitative evaluation we compute the macroscopic stress integral

$$\sigma_H^{\delta} = \int_{\Xi_H} \sigma_{\xi,H}^{\delta}$$

with

$$\sigma_{\xi,H}^{\delta} = \frac{1}{|\mathscr{Y}_{\xi}^{\delta}|} \int_{\mathscr{Y}_{\xi}^{\delta}} \mathbb{C}(\mathbf{y})[\boldsymbol{\varepsilon}_{\xi,H} + \boldsymbol{\varepsilon}(\mathbf{v}_{\xi,h})] \, d\mathbf{y}$$

depending on $\boldsymbol{\varepsilon}_{\xi,H} = \boldsymbol{\varepsilon}(\mathbf{u}_H)(\xi)$, see Table 4.1.

We observe, that the coarse resolution with dim $V_H = 765$ for the macro-problem is sufficient to approximate $|\sigma_H^{\delta}|$ with a relative error of less than 2%. On the micro-scale, a relative error of less than 10% requires at least dim $V_{\xi,h} = 6\,440\,067$ on $\mathscr{Y}_{\xi}^{\delta}$, and dim $V_{\xi,h} = 823\,875$ on $\mathscr{Y}_{\xi}^{\delta/2}$. The asymptotic results for $\mathscr{Y}_{\xi}^{\delta/4}$ show that in this case the resolution of the micro-structure is not sufficient to obtain approximations with a relative error smaller than 10%. For a detailed convergence analysis we refer to [24, Chap. 4].

Table 4.1 Numerical results for $|\sigma_H^\delta|$ computed on the macro-scale in V_H (row) and on the micro-scale in $V_{\xi,h}$ (column) for the micro-structures $\mathscr{Y}_\xi^\delta$, $\mathscr{Y}_\xi^{\delta/2}$, and $\mathscr{Y}_\xi^{\delta/4}$ with isotropic fibre distribution and 10% fibre volume fraction

DoFs	765	4 455	29 835	217 107	1 653 795
375	9.85312	9.81881	9.80574	9.80098	9.79927
2 187	8.63429	8.59902	8.58547	8.58050	8.57870
14 739	7.81376	7.77798	7.76413	7.75904	7.75720
107 811	5.83399	5.79482	5.77930	5.77350	5.77139
823 875	4.97443	4.93357	4.91718	4.91102	4.90876
6 440 067	4.60743	4.56594	4.54920	4.54287	4.54056
50 923 779	4.38415	4.34254	4.32568	4.31931	4.31697

DoFs	765	4 455	29 835	217 107	1 653 795
375	7.56932	7.53591	7.52304	7.51833	7.51663
2 187	7.97383	7.93835	7.92463	7.91960	7.91778
14 739	5.96454	5.92619	5.91105	5.90543	5.90339
107 811	5.06637	5.02650	5.01055	5.00456	5.00239
823 875	4.68158	4.64109	4.62479	4.61865	4.61641
6 440 067	4.50784	4.46708	4.45061	4.44440	4.44214
50 923 779	4.38927	4.34845	4.33192	4.32568	4.32341

DoFs	765	4 455	29 835	217 107	1 653 795
375	7.16115	7.12588	7.11215	7.10708	7.10525
2 187	5.52903	5.49100	5.47591	5.47028	5.46823
14 739	4.66845	4.62866	4.61271	4.60672	4.60454
107 811	4.29373	4.25314	4.23677	4.23059	4.22834
823 875	4.14409	4.10322	4.08669	4.08044	4.07816
6 440 067	4.04104	4.00008	3.98348	3.97720	3.97491
50 923 779	3.99777	3.95676	3.94012	3.93383	3.93153

Numerical tests for different fibre orientations and filler contents Now we investigate the elastic material properties for 10, 20 and 30% fibre volume content and $0°$, $45°$, $60°$ and $90°$ fibre orientation with respect to the applied load in the tensile test. We use a fine resolution in the RVE with dim $V_{\xi,h} = 6\,440\,067$, and dim $V_H = 6\,551\,523$ for the macro-solution. Each computation is performed on a single node[3] with 64 cores in approximately one hour.

The characteristic material values are evaluated by averaging in the cross section $\Omega_{\mathrm{ctr}} = (0, 0.5) \times (-0.2, 0.2) \times (0, 2) \subset \Omega$. Stress and strain in tensile direction are numerically computed by

---

[3] AMD Opteron 6376 processor with 2.3 GHz and 512 GB RAM.

Table 4.2 Characteristic material values for a uniaxial tensile test with unidirectional short fibres with 10, 20 and 30% fibre volume fraction and orientations between 0° and 90°

	PBT	0°	45°	60°	90°
E_z	2 500	6 116	3 333	3 167	3 338
ν_{zy}	0.40	0.38	0.44	0.36	0.21
ν_{zx}	0.40	0.38	0.36	0.41	0.51
σ_z	2.19371	5.30179	2.94166	2.78511	2.91609
ε_x	-3.50e-4	-3.32e-4	-3.15e-4	-3.56e-4	-4.47e-4
ε_y	-3.50e-4	-3.33e-4	-3.86e-4	-3.20e-4	-1.83e-4
ε_z	8.77e-4	8.66e-4	8.82e-4	8.79e-4	8.73e-4

	PBT	0°	45°	60°	90°
E_z	2 500	10 229	4 460	4 257	4 440
ν_{zy}	0.40	0.37	0.44	0.33	0.16
ν_{zx}	0.40	0.36	0.33	0.40	0.51
σ_z	2.19371	8.82239	3.94199	3.74542	3.8748
ε_x	-3.50e-4	-3.09e-4	-2.93e-4	-3.53e-4	-4.44e-4
ε_y	-3.50e-4	-3.19e-4	-3.89e-4	-2.92e-4	-1.40e-4
ε_z	8.77e-4	8.62e-4	8.83e-4	8.79e-4	8.72e-4

	PBT	0°	45°	60°	90°
E_z	2 500	15 443	6 621	6 032	5 973
ν_{zy}	0.40	0.34	0.39	0.31	0.13
ν_{zx}	0.40	0.34	0.33	0.38	0.50
σ_z	2.19371	13.293	5.83694	5.30212	5.21301
ε_x	-3.50e-4	-2.92e-4	-2.89e-4	-3.38e-4	-4.34e-4
ε_y	-3.50e-4	-2.93e-4	-3.45e-4	-2.68e-4	-1.15e-4
ε_z	8.77e-4	8.60e-4	8.81e-4	8.78e-4	8.72e-4

$$\sigma_z = \frac{1}{|\Omega_{\text{ctr}}|} \int_{\Xi_H \cap \Omega_{\text{ctr}}} \mathbb{C}_{\xi,H}[\boldsymbol{\varepsilon}(\mathbf{u}_H)]_{zz}, \quad \varepsilon_z = \frac{1}{|\Omega_{\text{ctr}}|} \int_{\Xi_H \cap \Omega_{\text{ctr}}} \boldsymbol{\varepsilon}(\mathbf{u}_H)_{zz} \quad (4.8)$$

to approximate Young's modulus $E_z = \frac{\sigma_z}{\varepsilon_z}$ in z-direction. Analogously, the strain averages ε_x and ε_y are defined to determine Poisson's ratios

$$\nu_{zx} = -\frac{\varepsilon_x}{\varepsilon_z}$$

and

$$v_{zy} = -\frac{\varepsilon_y}{\varepsilon_z},$$

cf. Table 4.2. All other orientations have a nearly equal stiffness character. Very accurate transverse isotropy, i.e. $v_{zx} = v_{zy}$, can be observed in the case of parallel fibre and tensile alignment. This corresponds to the symmetry of the rotational axis along the fibre direction. For each different fibre alignment to the acting force the material behaves strongly anisotropic. The full elastic properties of the multi-scale material are contained in the effective elasticity tensors

$$\mathbb{C}_{\xi,H}^{0°} = \begin{pmatrix} 6163.42 & 3976.58 & 3892.41 & 4.51 & -1.13 & 0.58 \\ 3976.58 & 6196.72 & 3906.77 & -0.17 & 0.47 & -0.44 \\ 3892.41 & 3906.77 & 9111.39 & 1.42 & 22.383 & 1.57 \\ 4.51 & -0.17 & 1.44 & 2219.04 & -0.62 & -0.71 \\ -1.13 & 0.47 & 22.38 & -0.62 & 2292.84 & 2.92 \\ 0.58 & -0.45 & 1.57 & -0.71 & 2.92 & 2253.73 \end{pmatrix},$$

$$\mathbb{C}_{\xi,H}^{45°} = \begin{pmatrix} 6189.56 & 3937.77 & 3936.05 & -7.96 & -43.42 & 13.42 \\ 3937.77 & 6886.21 & 4576.09 & 18.34 & 972.99 & 12.24 \\ 3936.05 & 4576.09 & 6897.25 & 25.26 & 977.66 & 27.34 \\ -7.96 & 18.35 & 25.26 & 2249.67 & 30.39 & 41.09 \\ -43.42 & 972.99 & 977.66 & 30.39 & 3642.93 & 29.39 \\ 13.42 & 12.24 & 27.34 & 41.09 & 29.39 & 2245.25 \end{pmatrix},$$

$$\mathbb{C}_{\xi,H}^{60°} = \begin{pmatrix} 6228.66 & 3938.61 & 3960.34 & -0.29 & -24.84 & 0.75 \\ 3938.61 & 7439.97 & 4308.68 & 13.46 & 972.10 & 4.09 \\ 3960.34 & 4308.68 & 6343.01 & 0.96 & 342.82 & -10.06 \\ -0.29 & 13.46 & 0.96 & 2303.76 & -0.30 & 42.83 \\ -24.84 & 972.10 & 342.82 & -0.30 & 3051.26 & 6.32 \\ 0.77 & 4.09 & -10.06 & 42.83 & 6.32 & 2246.48 \end{pmatrix},$$

$$\mathbb{C}_{\xi,H}^{90°} = \begin{pmatrix} 6163.42 & 3892.41 & 3976.58 & 0.58 & 1.13 & -4.51 \\ 3892.41 & 9111.39 & 3906.77 & 1.57 & -22.38 & -1.44 \\ 3976.58 & 3906.77 & 6196.72 & -0.45 & -0.47 & 0.17 \\ 0.58 & 1.57 & -0.45 & 2253.73 & -2.92 & 0.71 \\ 1.13 & -22.38 & -0.47 & -2.92 & 2292.84 & -0.62 \\ -4.51 & -1.44 & 0.17 & 0.71 & -0.62 & 2219.04 \end{pmatrix}.$$

The results fit well to experimental data provided by C. Röhrig [20, 21]. For a directly extruded specimen of type 1A with parallel aligned load fibre orientation and fibre volume content of 11.3% a Young's modulus of $E_z^{0°} = 5\,960$ and for 17.9% a modulus of $E_z^{0°} = 8\,240$ is measured. For fibre orientations with angle 0°, 45°, 60° and 90° measurements of Young's modulus of a specimen type 5A gives $E_z^{0°} = 6\,050$, $E_z^{45°} = 4\,300$, $E_z^{60°} = 3\,485$ and $E_z^{90°} = 3\,840$ for a volume filler content of 11.3%, and $E_z^{0°} = 7\,910$, $E_z^{45°} = 4\,500$, $E_z^{60°} = 4\,550$ and $E_z^{90°} = 4\,430$ for 17.3% fibre volume content.

4.4 Rate-independent Material Models with Memory

Inelastic effects can be described by material models with history variables. In this section we introduce a small-strain damage and plasticity model and the framework of energetic models as it is analysed in [15]. We specify the constitutive settings of these models, and we derive the return mapping and the corresponding consistent tangent operators.

Materials with memory We aim to find displacements $\mathbf{u}\colon [0, T] \times \Omega \to \mathbb{R}^3$ in the time interval $[0, T]$ of a material which is described by internal variables $\mathbf{z}\colon [0, T] \times \Omega \to \mathbb{R}^N$ and where the evolution is determined by the total energy and dissipation functionals

$$\mathcal{E}(t, \mathbf{u}, \mathbf{z}) = \int_\Omega W(\mathbf{x}, \boldsymbol{\varepsilon}(\mathbf{u}), \mathbf{z}) \, \mathrm{d}\mathbf{x} - \langle \ell(t), \mathbf{u} \rangle \,,$$

$$\mathcal{R}(\dot{\mathbf{z}}) = \int_\Omega R(\mathbf{x}, \dot{\mathbf{z}}) \, \mathrm{d}\mathbf{x} \,.$$

The load functional is given by

$$\langle \ell(t), \mathbf{u} \rangle = \int_\Omega \mathbf{b}(t) \cdot \mathbf{v} \, \mathrm{d}\mathbf{x} + \int_{\Gamma_\mathrm{N}} \mathbf{t}_\mathrm{N}(t) \cdot \mathbf{v} \, \mathrm{d}a$$

with body forces $\mathbf{b}\colon [0, T] \times \Omega \to \mathbb{R}^3$ and traction forces $\mathbf{t}_\mathrm{N}\colon [0, T] \times \Gamma_\mathrm{N} \to \mathbb{R}^3$.

We assume that the material is rate-independent, i.e., the inelastic deformation is independent from scaling in time. This is achieved if the dissipation function R is 1-homogeneous.

We only consider small strains with the ansatz space $V = \mathrm{H}^1(\Omega, \mathbb{R}^N)$ for the displacements, and the test space $V(\mathbf{0}) = \{\mathbf{v} \in V \colon \mathbf{v} = \mathbf{0} \text{ on } \Gamma_\mathrm{D}\}$ including homogeneous boundary conditions on the Dirichlet boundary Γ_D. For the internal variables we use the space $Z = \mathrm{L}_2(\Omega, \mathbb{R}^N)$. If the total energy functional $\mathcal{E}\colon [0, T] \times V \times Z \to \mathbb{R}$ is bounded and uniformly convex in $V(\mathbf{0}) \times Z$ for all $t \in [0, T]$, and if the dissipation functional $\mathcal{R}\colon Z \to \mathbb{R} \cup \{\infty\}$ is convex, proper and lower semi-continuous (l.s.c.), an energetic solution

$$(\mathbf{u}, \mathbf{z})\colon [0, T] \longrightarrow V \times Z$$

exists which is characterised by

$$\begin{aligned} \textbf{Equilibrium} \quad & 0 = \partial_\mathbf{u} \mathcal{E}\big(t, \mathbf{u}(t), \mathbf{z}(t)\big), \\ \textbf{Flow Rule} \quad & 0 \in \partial_\mathbf{z} \mathcal{E}(t, \mathbf{u}(t), \mathbf{z}(t)) + \partial \, \mathcal{R}\big(\dot{\mathbf{z}}(t)\big) \end{aligned}$$

and boundary conditions for the displacement $\mathbf{u}(t) = \mathbf{u}_\mathrm{D}(t)$ on the Dirichlet boundary Γ_D.

The incremental problem The evolution in time is approximated by a series of incremental problems. Let $0 = t_0 < t_1 < \cdots < t_{N_{\max}} = T$ be a time series with $\Delta t_n = t_n - t_{n-1}$. Starting with $(\mathbf{u}^0, \mathbf{z}^0)$ we define for $n = 1, \ldots, N_{\max}$ the following incremental problems depending on the given history variable $\mathbf{z}^{n-1}$, the load functional $\ell^n = \ell(t_n)$ and the Dirichlet data $\mathbf{u}_D^n = \mathbf{u}_D(t_n)$: find a minimiser $(\mathbf{u}^n, \mathbf{z}^n) \in V(\mathbf{u}_D^n) \times Z$ of the incremental functional

$$\mathscr{I}_n(\mathbf{u}^n, \mathbf{z}^n) = \mathscr{E}(t_n, \mathbf{u}^n, \mathbf{z}^n) + \mathscr{R}(\mathbf{z}^n - \mathbf{z}^{n-1}).$$

In our applications $\mathscr{I}_n(\cdot)$ is uniformly convex, so that a unique minimiser exists. It is determined by computing a critical point of $\mathscr{I}_n(\cdot)$ characterised by the nonlinear system

| **Equilibrium** | $0 = \partial_\mathbf{u}\mathscr{E}(t_n, \mathbf{u}^n, \mathbf{z}^n)$, | (4.9a) |
| **Flow Rule** | $0 \in \partial_\mathbf{z}\mathscr{E}(t_n, \mathbf{u}^n, \mathbf{z}^n) + \partial\mathscr{R}(\Delta\mathbf{z}^n)$. | (4.9b) |

Since we consider rate-independent materials, the flow rule is 1-homogeneous satisfying $\Delta t_n\mathscr{R}((\Delta t_n)^{-1}(\mathbf{z}^n - \mathbf{z}^{n-1})) = \mathscr{R}(\mathbf{z}^n - \mathbf{z}^{n-1})$ and thus depending only on the increment $\Delta\mathbf{z}^n = \mathbf{z}^n - \mathbf{z}^{n-1}$.

A simple damage model Continuum damage mechanics phenomenologically describes the expansion of micro-cracks and cavities with a single additional state variable

$$d = \frac{A_d}{A_0} \in [0, 1]$$

defined by the proportion of damaged area A_d of a representative cross sectional area A_0 (see, e.g., [9–11, 19]). This results into the effective stress response $\sigma = (1 - d)\mathbb{C}[\boldsymbol{\varepsilon}]$.

Within the energetic framework we set $N = 1$ and $\mathbf{z} = d$. The free energy is defined by

$$W(\boldsymbol{\varepsilon}, d) = (1 - d)W_{\text{elastic}}(\boldsymbol{\varepsilon}) + W_{\text{damage}}(d)$$

with $W_{\text{elastic}}(\boldsymbol{\varepsilon}) = \frac{1}{2}\boldsymbol{\varepsilon} : \mathbb{C}[\boldsymbol{\varepsilon}]$, so that $\sigma = \partial_\boldsymbol{\varepsilon} W(\boldsymbol{\varepsilon}, d) = (1 - d)\mathbb{C}[\boldsymbol{\varepsilon}]$. The dissipation functional

$$R_{\text{damage}}(\dot{d}) = \begin{cases} 0 & \dot{d} \geq 0, \\ +\infty & \text{otherwise} \end{cases} \tag{4.10}$$

just ensures the irreversibility of the damage process.

The additional term W_{damage} determines the evolution of the damage variable and guarantees $d < 1$. It is constructed in analogy to isotropic plasticity. We assume that the damage evolution is locally controlled by a strictly monotone function Φ

depending on the local elastic energy $Y = W_{\text{elastic}}(\boldsymbol{\varepsilon}(\mathbf{u}))$ and the complementarity conditions

$$\dot{d} \geq 0, \qquad \Phi(Y) - d \leq 0, \qquad (\Phi(Y) - d)\dot{d} = 0, \qquad (4.11)$$

i.e., d can only increase and not decrease and the material remains elastic, if the local elastic energy is small satisfying $\Phi(Y) < d$. Here, the choice

$$\Phi(Y) = 1 - \exp\left(-H(\sqrt{2Y} - Y_0)\right)$$

with damping and yielding point material parameters $H, Y_0 \geq 0$ is designed such that the material responds elastic for $\Phi(Y) < 0$, i.e., $\sqrt{2Y} < Y_0$, and $\Phi(Y) \longrightarrow 1$ for large Y. Nevertheless, since $\Phi(Y) < 1$ for all Y, the damage variable will not reach $d = 1$ which prevents fracture in this model.

Since $\Phi(\cdot)$ is assumed to be strictly monotone, the inverse is uniquely defined and the complementarity conditions (4.11) take the form

$$\dot{d} \geq 0, \qquad Y - \Phi^{-1}(d) \leq 0, \qquad (Y - \Phi^{-1}(d))\dot{d} = 0.$$

Due to our choice of the dissipation this is equivalent to

$$Y - \Phi^{-1}(d) \in \partial R_{\text{damage}}(\dot{d}).$$

This motivates the definition of the defect energy

$$W_{\text{damage}}(d) = \int_0^d \Phi^{-1}(\delta)\,d\delta, \qquad (4.12)$$

i.e., $\Phi^{-1}(d) = \partial_d W_{\text{damage}}(d)$ and $Y - \Phi^{-1}(d) = -\partial_d W(\boldsymbol{\varepsilon}, d)$.

For fixed $\boldsymbol{\varepsilon}$ the flow rule $0 \in \partial_d W(\boldsymbol{\varepsilon}, d) + \partial R_{\text{damage}}(\dot{d})$ characterises the minimum of $f(d; \boldsymbol{\varepsilon}) = W(\boldsymbol{\varepsilon}, d) + R_{\text{damage}}(\dot{d})$. Together, we observe that the minimiser is characterised by (4.11).

For our choice of Φ we observe $\Phi^{-1}(d) = \frac{1}{2}\left(Y_0 - \frac{1}{H}\log(1 - d)\right)^2$, so that $W_{\text{damage}}(d) \longrightarrow \infty$ for $d \longrightarrow 1$ prevents to reach a fully damaged material in this model. Nevertheless, we will see below that the inelastic energy is not uniformly convex, so that a well-defined evolution is only determined for sufficiently small loads or displacements.

Remark 4.4.1 This simple model does not allow to compute the transition to fracture. Due to the choice of $\Phi(\cdot)$, the energy in this model is only convex for sufficiency small damage, and the consistent tangent gets indefinite for d close to 1, so that one cannot expect convergence for the Newton method for large loads, and simulations for larger loads need an extension of the model.

The incremental flow rule for the damage model In the first step, we consider the semi-discrete problem in time. For given history variable d^{n-1}, the incremental problem (4.9) determines $(\mathbf{u}^n, d^n) \in V(\mathbf{u}_D^n) \times L_2(\Omega)$ with

$$0 = \int_{\Omega} \partial_{\varepsilon} W(\varepsilon(\mathbf{u}^n), d^n) : \varepsilon(\delta \mathbf{u}) \, d\mathbf{x} - \langle \ell^n, \delta \mathbf{u} \rangle, \qquad \delta \mathbf{u} \in V(0), \qquad (4.13a)$$

$$0 \in \partial_d W(\varepsilon(\mathbf{u}^n), d^n) + \partial R_{\text{damage}}(d^n - d^{n-1}). \qquad (4.13b)$$

The solution $(\mathbf{u}^n, d^n)$ of (4.13) is a critical point of the functional

$$J_n(\mathbf{u}, d) = \int_{\Omega} W(\varepsilon(\mathbf{u}), d) \, d\mathbf{x} - \langle \ell^n, \mathbf{u} \rangle + \int_{\Omega} R_{\text{damage}}(d - d^{n-1}) \, d\mathbf{x} \qquad (4.14)$$

subject to the essential boundary conditions $\mathbf{u} = \mathbf{u}_D^n$ on Γ_D.

The incremental damage problem can be reduced to a nonlinear problem for the displacement by inserting the local solution of the incremental flow rule depending on the strain. This is based on the following result.

Lemma 4.2 *For given history variable d^{n-1} and fixed strain ε the unique solution Δd of the local incremental flow rule in every material point*

$$0 \in \partial_d W(\varepsilon, d^{n-1} + \Delta d) + \partial R_{damage}(\Delta d)$$

is given by

$$\Delta d = \max \left\{ 0, \, \Phi(Y(\varepsilon)) - d^{n-1} \right\}, \qquad Y(\varepsilon) = W_{elastic}(\varepsilon).$$

Proof Evaluating $\partial_d W(\varepsilon, d^{n-1} + \Delta d) = -Y(\varepsilon) + \Phi^{-1}(d^{n-1} + \Delta d)$ in the incremental flow rule yields

$$Y(\varepsilon) - \Phi^{-1}(d^{n-1} + \Delta d) \in \partial R_{\text{damage}}(\Delta d) = \begin{cases} \{0\} & \Delta d > 0, \\ (-\infty, 0] & \Delta d = 0, \\ \emptyset & \Delta d < 0. \end{cases}$$

This is equivalent to the complementarity condition

$$\Delta d \geq 0, \quad \Phi(Y(\varepsilon)) - d^{n-1} - \Delta d \leq 0, \quad \left(\Phi(Y(\varepsilon)) - d^{n-1} - \Delta d\right)\Delta d = 0$$

which directly implies $\Delta d = \max \left\{ 0, \, \Phi(Y(\varepsilon)) - d^{n-1} \right\}$. □

The evaluation of the flow rule defines the update of the damage variable

$$d_n(\varepsilon) = d^{n-1} + \Delta d = d^{n-1} + \max \left\{ 0, \, \Phi(Y(\varepsilon)) - d^{n-1} \right\}$$

and the incremental stress response

$$\sigma_n(\varepsilon) = \left(1 - d_n(\varepsilon)\right)\mathbb{C}[\varepsilon].$$

Choosing $\text{sgn}(s) \in \partial \max\{0, s\}$ with

$$\text{sgn}(s) = \begin{cases} 1 & s > 0, \\ 0 & \text{otherwise} \end{cases}$$

defines the consistent tangent $\mathbb{C}_n(\varepsilon) \in \partial \sigma_n(\varepsilon)$ by

$$\mathbb{C}_n(\varepsilon)[\delta\varepsilon] = \left(1 - d_n(\varepsilon)\right)\mathbb{C}(\varepsilon)[\delta\varepsilon]$$
$$- \text{sgn}\left(\max\left\{0, \Phi(Y(\varepsilon)) - d^{n-1}\right\}\right)\Phi'(Y(\varepsilon))\left(\mathbb{C}[\varepsilon] \cdot \delta\varepsilon\right)\mathbb{C}[\varepsilon]$$

with $\Phi'(Y) = \frac{H}{\sqrt{2Y}}\exp(-H(\sqrt{2Y} - Y_0))$.

The incremental problem can be solved by minimising the reduced functional for the displacement

$$J_n^{\text{red}}(\mathbf{u}) = \int_\Omega \left(\int_0^{Y(\varepsilon(\mathbf{u}))}\left(1 - d^{n-1} - \max\left\{0, \Phi(y) - d^{n-1}\right\}\right) dy\right) dx - \langle \ell^n, \mathbf{u}\rangle$$

with first variation

$$\partial J_n^{\text{red}}(\mathbf{u})[\delta\mathbf{u}] =$$
$$\int_\Omega \left(1 - d^{n-1} - \max\left\{0, \Phi(Y(\varepsilon(\mathbf{u}))) - d^{n-1}\right\}\right)\mathbb{C}(\varepsilon(\mathbf{u}))[\varepsilon(\delta\mathbf{u})]\,dx - \langle \ell^n, \delta\mathbf{u}\rangle$$

corresponding to (4.13a). Within a generalised Newton method, the consistent tangent defines

$$\partial^2 J_n^{\text{red}}(\mathbf{u})[\Delta\mathbf{u}, \delta\mathbf{u}] = \int_\Omega \mathbb{C}_n(\varepsilon(\mathbf{u}))[\varepsilon(\Delta\mathbf{u})] : \varepsilon(\delta\mathbf{u})\,dx\,.$$

We observe $\mathbb{C}_n(\varepsilon)[\varepsilon] : \varepsilon < 0$ for large ε, so that for our choice of Φ the second variation of J_n^{red} is not positive and thus J_n^{red} is only convex for sufficiently small strains. This restricts the application of our damage model to moderate loads. An extended damage model with convex energy can be obtained by including gradient terms, see, e.g., [29].

Small strain elasto-plasticity For the elasto-plastic model with hardening, the internal variables $\mathbf{z} = (\varepsilon_p, r)$ are the plastic strain ε_p with $\text{tr}\varepsilon_p = 0$ and the isotropic hardening parameter r, i.e., $N = 6$. The strain is decomposed into elastic and plastic part $\varepsilon(\mathbf{u}) = \varepsilon_e + \varepsilon_p$, and the free energy is given by

$$W(\mathbf{x}, \varepsilon, \varepsilon_p, r) = W_{\text{elastic}}(\mathbf{x}, \varepsilon(\mathbf{u}) - \varepsilon_p) + W_{\text{plastic}}(\varepsilon_p, r)$$

with the elastic energy (4.1) and defect energy

$$W_{\text{plastic}}(\boldsymbol{\varepsilon}_{\text{p}}, r) = W_{\text{kin}}(\boldsymbol{\varepsilon}_{\text{p}}) + W_{\text{iso}}(r) \tag{4.15}$$

combining kinematic and isotropic hardening. The translation of the yield surface is described by kinematic hardening with

$$W_{\text{kin}}(\boldsymbol{\varepsilon}_{\text{p}}) = \frac{1}{2} K \boldsymbol{\varepsilon}_{\text{p}} : \boldsymbol{\varepsilon}_{\text{p}}$$

depending on the hardening parameter $K \geq 0$. The expansion of the yield surface is described by isotropic hardening determined by the yield function

$$\Psi(r) = \sigma_{\text{y}} + H_0 r + (K_\infty - K_0)(1 - \exp(\delta r))$$

for given material parameters σ_{y}, H_0, $\delta \geq 0$ and $K_\infty \geq K_0 \geq 0$. Now we construct the remaining energy contribution and the dissipation such that the plastic evolution satisfies the yield condition $|\operatorname{dev}\sigma - \beta| + \Psi(r) \leq 0$.

The energy definition corresponds to the constitutive stress-stain relation

$$\sigma = \partial_{\boldsymbol{\varepsilon}} W(\boldsymbol{\varepsilon}, \boldsymbol{\varepsilon}_{\text{p}}, r) = \partial_{\boldsymbol{\varepsilon}} W_{\text{elastic}}(\boldsymbol{\varepsilon}(\mathbf{u}) - \boldsymbol{\varepsilon}_{\text{p}}) = \mathbb{C}[\boldsymbol{\varepsilon}(\mathbf{u}) - \boldsymbol{\varepsilon}_{\text{p}}].$$

It defines the back stress $\beta = \partial_{\boldsymbol{\varepsilon}_{\text{p}}} W_{\text{kin}}(\boldsymbol{\varepsilon}_e) = K \boldsymbol{\varepsilon}_{\text{p}}$, and the conjugated variables $\mathbf{y} = (\alpha, \zeta) = -\partial_{\mathbf{z}} W(\boldsymbol{\varepsilon}, \mathbf{z})$ with

$$\alpha = \partial_{\boldsymbol{\varepsilon}_{\text{p}}} W_{\text{elastic}}(\boldsymbol{\varepsilon}(\mathbf{u}) - \boldsymbol{\varepsilon}_{\text{p}}) - \partial_{\boldsymbol{\varepsilon}_{\text{p}}} W_{\text{kin}}(\boldsymbol{\varepsilon}_{\text{p}}) = \operatorname{dev}\sigma - \beta,$$
$$\zeta = -\partial_r W_{\text{iso}}(r).$$

This yields the constitutive relation $\zeta = -\partial_r W_{\text{iso}}(r) = -\Psi(r)$ by defining

$$W_{\text{iso}}(r) = \int_0^r \Psi(\rho)\, d\rho.$$

The plastic evolution is determined by the plastic potential

$$R^*_{\text{plastic}}(\alpha, \zeta) = \begin{cases} 0 & |\alpha| + \zeta \leq 0 \text{ and } \zeta \leq 0, \\ +\infty & \text{otherwise}, \end{cases}$$

which is by duality equivalent to the dissipation functional

$$R_{\text{plastic}}(\dot{\boldsymbol{\varepsilon}}_{\text{p}}, \dot{r}) = \begin{cases} 0 & \dot{r} \geq |\dot{\boldsymbol{\varepsilon}}_{\text{p}}|, \\ +\infty & \text{otherwise}. \end{cases} \tag{4.16}$$

The flow rule $(\alpha, \zeta) \in \partial R_{\text{plastic}}(\dot{\boldsymbol{\varepsilon}}_{\text{p}}, \dot{r})$ in every material point is evaluated by duality, i.e., $(\dot{\boldsymbol{\varepsilon}}_{\text{p}}, \dot{r}) \in \partial R^*_{\text{plastic}}(\alpha, \zeta)$. Introducing a consistency parameter λ_{p} this is equivalent to the normality rule

$$\dot{\varepsilon}_p = \lambda_p \frac{\alpha}{|\alpha|}, \qquad \dot{r} = \lambda_p,$$

and the complementarity conditions

$$\lambda_p \geq 0, \qquad |\alpha| + \zeta \leq 0, \qquad \lambda_p(|\alpha| + \zeta) = 0.$$

In particular, this implies $|\dot{\varepsilon}_p| = \dot{r}$, and assuming $\varepsilon_p(0) = 0$ and $r(0) = 0$ at initial time $t = 0$, we obtain

$$r(t) = \int_0^t \dot{r} \, dt = \int_0^t |\dot{\varepsilon}_p| \, dt,$$

i.e., r is the equivalent plastic strain.

The incremental flow rule for elasto-plasticity For the incremental problem the local computation of the stress response and the consistent tangent in every material point in the RVEs is reduced to a scalar nonlinear problem for the equivalent plastic strain increment.

Lemma 4.3 *For given history variables* $(\varepsilon_p^{n-1}, r^{n-1})$ *and strain* ε *a unique solution* $(\Delta \varepsilon_p, \Delta r)$ *of the local flow rule in every material point*

$$0 \in \partial_{(\varepsilon_p, r)} W(\varepsilon, \varepsilon_p^{n-1} + \Delta \varepsilon_p, r^{n-1} + \Delta r) + \partial \, R_{plastic}(\Delta \varepsilon_p, \Delta r)$$

exists.

Proof For fixed history and given strain ε, the increment is a minimiser of

$$\Phi(\Delta \varepsilon_p, \Delta r; \varepsilon) = W(\varepsilon, \varepsilon_p^{n-1} + \Delta \varepsilon_p, r^{n-1} + \Delta r) + R_{plastic}(\Delta \varepsilon_p, \Delta r).$$

Since the functional is uniformly convex in $\mathbb{R}^6$, the minimiser exists and it is unique. For the evaluation we define the conjugated variables

$$(\alpha, \zeta) = -\partial_{(\varepsilon_p, r)} W(\varepsilon, \varepsilon_p^{n-1} + \Delta \varepsilon_p, r^{n-1} + \Delta r).$$

We obtain $\alpha = \mathrm{dev}\,\sigma - K(\varepsilon_p^{n-1} + \Delta \varepsilon_p)$ from the stress $\sigma = \mathbb{C}[\varepsilon - \varepsilon_p^{n-1} - \Delta \varepsilon_p]$, and $\zeta = -\Psi(r^{n-1} + \Delta r)$. Evaluating

$$(\Delta \varepsilon_p, \Delta r) \in \partial \, R^*_{plastic}(\alpha, \zeta) = \begin{cases} \{0\} & |\alpha| + \zeta < 0, \\ [0, \infty) \begin{pmatrix} \frac{\alpha}{|\alpha|} \\ 1 \end{pmatrix} & |\alpha| + \zeta = 0, \\ \emptyset & |\alpha| + \zeta > 0 \end{cases}$$

yields the normality rule

$$\begin{pmatrix} \triangle\, \boldsymbol{\varepsilon}_{\mathrm{p}} \\ \triangle\, r \end{pmatrix} = \lambda_{\mathrm{p}} \begin{pmatrix} \frac{\alpha}{|\alpha|} \\ 1 \end{pmatrix}$$

and the complementarity conditions for the consistency parameter λ_{p}

$$\lambda_{\mathrm{p}} \geq 0\,, \quad |\alpha| + \zeta \leq 0\,, \quad \lambda_{\mathrm{p}}(|\alpha| + \zeta) = 0\,.$$

The normality rule yields $\triangle r = \lambda_{\mathrm{p}} = |\triangle\boldsymbol{\varepsilon}_{\mathrm{p}}|$ and for the flow direction

$$\frac{\triangle\boldsymbol{\varepsilon}}{|\triangle\boldsymbol{\varepsilon}|} = \frac{\alpha}{|\alpha|} = \frac{2\mu\,\mathrm{dev}\,\boldsymbol{\varepsilon} - (2\mu + K)(\boldsymbol{\varepsilon}_{\mathrm{p}}^{n-1} + \triangle\boldsymbol{\varepsilon}_{\mathrm{p}})}{|2\mu\,\mathrm{dev}\,\boldsymbol{\varepsilon} - (2\mu + K)(\boldsymbol{\varepsilon}_{\mathrm{p}}^{n-1} + \triangle\boldsymbol{\varepsilon}_{\mathrm{p}})|} = \frac{\alpha_n^{\mathrm{tr}}(\boldsymbol{\varepsilon})}{|\alpha_n^{\mathrm{tr}}(\boldsymbol{\varepsilon})|}$$

with the relative trial stress $\alpha_n^{\mathrm{tr}}(\boldsymbol{\varepsilon}) = 2\mu\,\mathrm{dev}\,\boldsymbol{\varepsilon} - (2\mu + K)\boldsymbol{\varepsilon}_{\mathrm{p}}^{n-1}$. Thus, defining the flow function

$$F_n(\triangle r; \boldsymbol{\varepsilon}) = |\alpha_n^{\mathrm{tr}}(\boldsymbol{\varepsilon})| - (2\mu + K)\triangle r - \Psi(r^{n-1} + \triangle r)$$

we observe $|\alpha| + \zeta = F_n(\triangle r; \boldsymbol{\varepsilon})$. Now, for the given strain $\boldsymbol{\varepsilon}$ we have to distinguish two cases. If $F_n(0; \boldsymbol{\varepsilon}) \leq 0$, we set $\triangle r = 0$. Otherwise, since $F_n(\cdot; \boldsymbol{\varepsilon})$ is strictly monotone and negative for large $\triangle r$, the equation $F_n(\triangle r; \boldsymbol{\varepsilon}) = 0$ uniquely determines $\triangle r > 0$. Then, we obtain

$$\triangle\boldsymbol{\varepsilon}_{\mathrm{p}} = \triangle r\, \frac{\alpha_n^{\mathrm{tr}}(\boldsymbol{\varepsilon})}{|\alpha_n^{\mathrm{tr}}(\boldsymbol{\varepsilon})|}\,. \qquad\qquad \square$$

Evaluating the increment $\triangle r_n(\boldsymbol{\varepsilon})$ defines the update of the history variables

$$r_n(\boldsymbol{\varepsilon}) = r^{n-1} + \triangle r_n(\boldsymbol{\varepsilon})\,,$$

$$\boldsymbol{\varepsilon}_{\mathrm{p},n}(\boldsymbol{\varepsilon}) = \boldsymbol{\varepsilon}_{\mathrm{p}}^{n-1} + \triangle r_n(\boldsymbol{\varepsilon})\frac{\alpha_n^{\mathrm{tr}}(\boldsymbol{\varepsilon})}{|\alpha_n^{\mathrm{tr}}(\boldsymbol{\varepsilon})|}\,,$$

the incremental stress response

$$\sigma_n(\boldsymbol{\varepsilon}) = \mathbb{C}[\boldsymbol{\varepsilon} - \boldsymbol{\varepsilon}_{\mathrm{p},n}(\boldsymbol{\varepsilon})]\,,$$

and the consistent tangent $\mathbb{C}_n(\boldsymbol{\varepsilon}) \in \partial\sigma_n(\boldsymbol{\varepsilon})$ by

$$\mathbb{C}_n(\boldsymbol{\varepsilon})[\delta\boldsymbol{\varepsilon}] = \mathbb{C}[\delta\boldsymbol{\varepsilon}] - \frac{4\mu^2\triangle r_n(\boldsymbol{\varepsilon})}{|\alpha_n^{\mathrm{tr}}(\boldsymbol{\varepsilon})|}\,\mathrm{dev}(\delta\boldsymbol{\varepsilon})$$

$$+ \left(\frac{4\mu^2\triangle r_n(\boldsymbol{\varepsilon})}{|\alpha_n^{\mathrm{tr}}(\boldsymbol{\varepsilon})|} - \frac{4\mu^2}{2\mu + K + \Psi'(r_n(\boldsymbol{\varepsilon}))} \right) \frac{\alpha_n^{\mathrm{tr}}(\boldsymbol{\varepsilon})\cdot\delta\boldsymbol{\varepsilon}}{|\alpha_n^{\mathrm{tr}}(\boldsymbol{\varepsilon})|} \frac{\alpha_n^{\mathrm{tr}}(\boldsymbol{\varepsilon})}{|\alpha_n^{\mathrm{tr}}(\boldsymbol{\varepsilon})|}\,,$$

cf. [25, Sect. 3.3.2].

Damage and elasto-plasticity For the model combining damage and elasto-plasticity we use the internal variables $\mathbf{z} = (d, \boldsymbol{\varepsilon}_p, r)$ with $N = 7$ components and the free energy

$$W(\mathbf{x}, \boldsymbol{\varepsilon}, \mathbf{z}) = (1 - d)W_{\text{elastic}}(\mathbf{x}, \boldsymbol{\varepsilon}(\mathbf{u}) - \boldsymbol{\varepsilon}_p) + W_{\text{defect}}(\mathbf{z})$$
$$= (1 - d)\big(W_{\text{elastic}}(\mathbf{x}, \boldsymbol{\varepsilon}(\mathbf{u}) - \boldsymbol{\varepsilon}_p) + W_{\text{plastic}}(\boldsymbol{\varepsilon}_p, r)\big) + W_{\text{damage}}(d)$$

with the elastic energy (4.1) and defect energy components (4.12) and (4.15). The dissipation combines (4.10) and (4.16) to

$$\mathscr{R}(\dot{d}, \dot{\boldsymbol{\varepsilon}}_p, \dot{r}) = \int_\Omega R(\dot{d}, \dot{\boldsymbol{\varepsilon}}_p, \dot{r})\, d\mathbf{x}, \quad R(\dot{d}, \dot{\boldsymbol{\varepsilon}}_p, \dot{r}) = R_{\text{damage}}(\dot{d}) + R_{\text{plastic}}(\dot{\boldsymbol{\varepsilon}}_p, \dot{r}).$$

The incremental two-scale elasto-plastic damage model For the incremental problem the local computation of the stress response and the consistent tangent in every material point in the RVEs is evaluated first for the plasticity variables and then for the damage variable.

Lemma 4.4 *For given history variables $(d^{n-1}, \boldsymbol{\varepsilon}_p^{n-1}, r^{n-1})$ and strain $\boldsymbol{\varepsilon}$ a unique solution $(\Delta d, \Delta \boldsymbol{\varepsilon}_p, \Delta r)$ of the local flow rule in every material point*

$$0 \in \partial_{(d, \boldsymbol{\varepsilon}_p, r)} W(\boldsymbol{\varepsilon}, d^{n-1} + \Delta d, \boldsymbol{\varepsilon}_p^{n-1} + \Delta \boldsymbol{\varepsilon}_p, r^{n-1} + \Delta r) + \partial\, R(\Delta d, \Delta \boldsymbol{\varepsilon}_p, \Delta r)$$

exists.

Proof Inserting the conjugated variables

$$- \partial_{(d, \boldsymbol{\varepsilon}_p, r)} W(\boldsymbol{\varepsilon}, d^{n-1} + \Delta d, \boldsymbol{\varepsilon}_p^{n-1} + \Delta \boldsymbol{\varepsilon}_p, r^{n-1} + \Delta r)$$
$$= \begin{pmatrix} Y - \Phi^{-1}(d^{n-1} + \Delta d) \\ (1 - d^{n-1} + \Delta d)\big(\operatorname{dev}(\sigma) - K(\boldsymbol{\varepsilon}_p^{n-1} + \Delta \boldsymbol{\varepsilon}_p)\big) \\ -(1 - d^{n-1} + \Delta d)\Psi(r^{n-1} + \Delta r) \end{pmatrix} = \begin{pmatrix} Y - \Phi^{-1}(d^{n-1} + \Delta d) \\ (1 - d^{n-1} + \Delta d)\alpha \\ (1 - d^{n-1} + \Delta d)\zeta \end{pmatrix}$$

with $Y = W_{\text{elastic}}(\boldsymbol{\varepsilon} - \boldsymbol{\varepsilon}_p^{n-1} - \Delta \boldsymbol{\varepsilon}_p)$, the back stress $\alpha = \operatorname{dev} \sigma - K(\boldsymbol{\varepsilon}_p^{n-1} + \Delta \boldsymbol{\varepsilon}_p)$, $\zeta = -\Psi(r^{n-1} + \Delta r)$, and $\sigma = \mathbb{C}[\boldsymbol{\varepsilon} - \boldsymbol{\varepsilon}_p^{n-1} - \Delta \boldsymbol{\varepsilon}_p]$ into the flow rule yields

$$Y - \Phi^{-1}(d^{n-1} + \Delta d) \in \partial\, R_{\text{damage}}(\Delta d),$$

and, since $R_{\text{plastic}}^*(\cdot)$ is 1-homogeneous,

$$(\Delta \boldsymbol{\varepsilon}_p, \Delta r) \in \partial\, R_{\text{plastic}}^*(\alpha, \zeta) = \partial\, R_{\text{plastic}}^*\Big((1 - d^{n-1} + \Delta d)(\alpha, \zeta)\Big).$$

This shows that in the first step, the plastic increment can be evaluated from the plastic flow rule independent from the damage variable. We proceed as in Lemma 4.3. The plastic flow rule is equivalent to the normality rule

$$\begin{pmatrix} \triangle \boldsymbol{\varepsilon}_p \\ \triangle r \end{pmatrix} = \lambda_p \begin{pmatrix} \frac{\alpha}{|\alpha|} \\ 1 \end{pmatrix}$$

and the complementarity conditions for the consistency parameter λ_p

$$\lambda_p \geq 0, \quad |\alpha| + \zeta \leq 0, \quad \lambda_p(|\alpha| + \zeta) = 0.$$

The normality rule yields $\triangle r = \lambda_p = |\triangle \boldsymbol{\varepsilon}_p|$ and for the flow direction

$$\frac{\triangle \boldsymbol{\varepsilon}}{|\triangle \boldsymbol{\varepsilon}|} = \frac{\alpha}{|\alpha|} = \frac{2\mu \operatorname{dev} \boldsymbol{\varepsilon} - (2\mu + K)(\boldsymbol{\varepsilon}_p^{n-1} + \triangle \boldsymbol{\varepsilon}_p)}{|2\mu \operatorname{dev} \boldsymbol{\varepsilon} - (2\mu + K)(\boldsymbol{\varepsilon}_p^{n-1} + \triangle \boldsymbol{\varepsilon}_p)|} = \frac{\alpha_n^{tr}(\boldsymbol{\varepsilon})}{|\alpha_n^{tr}(\boldsymbol{\varepsilon})|}$$

with the relative trial stress $\alpha_n^{tr}(\boldsymbol{\varepsilon}) = 2\mu \operatorname{dev} \boldsymbol{\varepsilon} - (2\mu + K)\boldsymbol{\varepsilon}_p^{n-1}$. Thus, defining

$$F_n(\triangle r; \boldsymbol{\varepsilon}) = |\alpha_n^{tr}(\boldsymbol{\varepsilon})| - (2\mu + K)\triangle r - \Psi(r^{n-1} + \triangle r)$$

we observe $|\alpha| + \zeta = F_n(\triangle r; \boldsymbol{\varepsilon})$. Now, for the given $\boldsymbol{\varepsilon}$ we have to distinguish two cases. If $F_n(0; \boldsymbol{\varepsilon}) \leq 0$, we set $\triangle r = 0$. Otherwise, $\triangle r > 0$ is uniquely determined by the equation $F_n(\triangle r; \boldsymbol{\varepsilon}) = 0$. Then, we obtain

$$\triangle \boldsymbol{\varepsilon}_p = \triangle r \frac{\alpha_n^{tr}(\boldsymbol{\varepsilon})}{|\alpha_n^{tr}(\boldsymbol{\varepsilon})|}$$

which also defines $Y_n(\boldsymbol{\varepsilon}) = W_{\text{elastic}}(\boldsymbol{\varepsilon} - \boldsymbol{\varepsilon}_p^{n-1} - \triangle \boldsymbol{\varepsilon}_p)$. Now, the increment of the damage variable is computed as in Lemma 4.2 depending on $Y_n(\boldsymbol{\varepsilon})$, i.e.,

$$\triangle d = \max \left\{ 0, \Phi(Y_n(\boldsymbol{\varepsilon})) - d^{n-1} \right\}.$$

$\square$

The evaluation of the flow rule defines the update of the history variables

$$d_n(\boldsymbol{\varepsilon}) = d^{n-1} + \triangle d,$$
$$\boldsymbol{\varepsilon}_{p,n}(\boldsymbol{\varepsilon}) = \boldsymbol{\varepsilon}_p^{n-1} + \triangle \boldsymbol{\varepsilon}_p,$$
$$r_n(\boldsymbol{\varepsilon}) = r^{n-1} + \triangle r,$$

the incremental stress response $\sigma_n(\boldsymbol{\varepsilon}) = (1 - d_n(\boldsymbol{\varepsilon}))\mathbb{C}[\boldsymbol{\varepsilon} - \boldsymbol{\varepsilon}_{p,n}(\boldsymbol{\varepsilon})]$, and the consistent tangent $\mathbb{C}_n(\boldsymbol{\varepsilon}) \in \partial \sigma_n(\boldsymbol{\varepsilon})$ with

$$\mathbb{C}_n(\boldsymbol{\varepsilon})[\delta \boldsymbol{\varepsilon}] = (1 - d_n(\boldsymbol{\varepsilon}))\mathbb{C}_n^{\text{plastic}}(\boldsymbol{\varepsilon})[\delta \boldsymbol{\varepsilon}]$$
$$- \operatorname{sgn}\left(\max \left\{ 0, \Phi(Y_n(\boldsymbol{\varepsilon})) - d^{n-1} \right\}\right)\left(\mathbb{C}[\boldsymbol{\varepsilon} - \boldsymbol{\varepsilon}_{p,n}(\boldsymbol{\varepsilon})] \cdot \delta \boldsymbol{\varepsilon}\right)\mathbb{C}[\boldsymbol{\varepsilon}]$$

and

$$
\mathbb{C}_n^{\text{plastic}}(\boldsymbol{\varepsilon})[\delta\boldsymbol{\varepsilon}] = \mathbb{C}[\delta\boldsymbol{\varepsilon}] - \frac{4\mu^2\Delta r}{|\alpha_n^{\text{tr}}(\boldsymbol{\varepsilon})|}\, \text{dev}(\delta\boldsymbol{\varepsilon})
$$
$$
+ \left(\frac{4\mu^2\Delta r}{|\alpha_n^{\text{tr}}(\boldsymbol{\varepsilon})|} - \frac{4\mu^2}{2\mu + K + \Psi'(r_n(\boldsymbol{\varepsilon}))}\right) \frac{\alpha_n^{\text{tr}}(\boldsymbol{\varepsilon}) \cdot \delta\boldsymbol{\varepsilon}}{|\alpha_n^{\text{tr}}(\boldsymbol{\varepsilon})|} \frac{\alpha_n^{\text{tr}}(\boldsymbol{\varepsilon})}{|\alpha_n^{\text{tr}}(\boldsymbol{\varepsilon})|} .
$$

4.5 Heterogeneous Two-Scale FEM for Inelasticity

The inelastic material behaviour for short fibre reinforced polymers is modelled by a two-scale infinitesimal elasto-plastic damage material [27, 28]. Here, this model is reformulated in the framework of generalised standard materials which directly defines the corresponding algorithmic realisation within the FE^2 framework.

Two-scale models with memory The energetic framework extends to the two-scale setting as follows. We consider $\mathbf{u}_H \colon [0,T] \to V_H$ on the macro-scale satisfying the Dirichlet boundary conditions, i.e., $\mathbf{u}_H(t) \in V_H(\mathbf{u}_D(t))$, and locally in every RVE $\mathscr{Y}_\xi$ the micro-fluctuations $\mathbf{v}_{\xi,h} \colon [0,T] \to V_{\xi,h}$ and the internal variables describing the material history $\mathbf{z}_{\xi,h} \colon [0,T] \to Z_{\xi,h}$ has to be determined. Here, we use for the internal variables piecewise constant vectors in $Z_{\xi,h} \subset L_2(\mathscr{Y}_\xi, \mathbb{R}^N)$ represented by $\mathbf{z}_{\xi,h}(\zeta) \in \mathbb{R}^N$ at the integration points $\zeta \in \varXi_{\xi,h} \subset \mathscr{Y}_\xi$ in the RVE. Together, we define $Z_h = \prod_{\xi \in \varXi_H} Z_{\xi,h}$.

The model is determined by the corresponding two-scale energy and dissipation functionals

$$
\mathscr{E}_H(t, \mathbf{u}_H, \mathbf{v}_h, \mathbf{z}_h) = \int_{\varXi_H} W_\xi(\boldsymbol{\varepsilon}(\mathbf{u}_H), \mathbf{v}_{\xi,h}, \mathbf{z}_{\xi,h}) - \langle \ell(t), \mathbf{u}_H \rangle ,
$$
$$
\mathscr{R}_H(\dot{\mathbf{z}}_h) = \int_{\varXi_H} R_\xi(\dot{\mathbf{z}}_{\xi,h}) ,
$$

where the contributions at every sample point $\xi \in \varXi_H$ is evaluated in the RVEs $\mathscr{Y}_\xi$ by the locally averaged two-scale micro-energy and micro-dissipation

$$
W_\xi(\boldsymbol{\varepsilon}_H, \mathbf{v}_{\xi,h}, \mathbf{z}_{\xi,h}) = \frac{1}{|\mathscr{Y}_\xi|} \int_{\mathscr{Y}_\xi} W(\mathbf{x}, \boldsymbol{\varepsilon}_{\xi,H} + \boldsymbol{\varepsilon}(\mathbf{v}_{\xi,h}), \mathbf{z}_{\xi,h})\, d\mathbf{x} ,
$$
$$
R_\xi(\dot{\mathbf{z}}_h) = \frac{1}{|\mathscr{Y}_\xi|} \int_{\mathscr{Y}_\xi} R(\mathbf{x}, \dot{\mathbf{z}}_{\xi,h})\, d\mathbf{x} ,
$$

depending on the macro-strain $\boldsymbol{\varepsilon}_{\xi,H} = \boldsymbol{\varepsilon}(\mathbf{u}_H)(\xi)$. Again, this defines the micro-solution by $\mathbf{u}_{\xi,h} = \mathbf{u}_{\xi,H} + \mathbf{v}_{\xi,h}$ depending on the linearised macro-solution by $\mathbf{u}_{\xi,H}(\mathbf{x}) = \mathbf{u}_H(\xi) + D\mathbf{u}_H(\xi)(\mathbf{x} - \xi)$, i.e., by construction the micro-fluctuation is periodic and the strain of the macro-solution $\boldsymbol{\varepsilon}(\mathbf{u}_{\xi,H}) \equiv \boldsymbol{\varepsilon}_{\xi,H}$ is constant in $\mathscr{Y}_\xi$.

The incremental two-scale problem Starting with $(\mathbf{u}_H^0, \mathbf{v}_h^0, \mathbf{z}_h^0)$ we solve for $n = 1, \ldots, N_{\max}$ the following incremental problems depending on the given history variable $\mathbf{z}_h^{n-1}$: find a minimiser $(\mathbf{u}_H^n, \mathbf{v}_h^n, \mathbf{z}_h^n) \in V_H(\mathbf{u}_D^n) \times V_h \times Z_h$ of the two-scale incremental functional

$$\mathscr{J}_{h,n}(\mathbf{u}_H^n, \mathbf{v}_h^n, \mathbf{z}_h^n) = \mathscr{E}_H(t_n, \mathbf{u}_H^n, \mathbf{v}_h^n, \mathbf{z}_h^n) + \mathscr{R}_H(\mathbf{z}_h^n - \mathbf{z}_h^{n-1}) \,.$$

The minimiser is determined by computing a critical point of $\mathscr{J}_{h,n}(\cdot)$. This is characterised by the nonlinear system

Macro-Equilibrium	$0 = \partial_{\mathbf{u}} \mathscr{E}_H(t_n, \mathbf{u}_H^n, \mathbf{v}_h^n, \mathbf{z}_h^n) \,,$	(4.17a)
Micro-Equilibrium	$0 = \partial_{\mathbf{v}} \mathscr{E}_H(t_n, \mathbf{u}_H^n, \mathbf{v}_h^n, \mathbf{z}_h^n) \,,$	(4.17b)
Flow Rule	$0 \in \partial_{\mathbf{z}} \mathscr{E}_H(t_n, \mathbf{u}_H^n, \mathbf{v}_h^n, \mathbf{z}_h^n) + \partial \, \mathscr{R}_H(\triangle \mathbf{z}_h^n) \,.$	(4.17c)

The **Macro-Equilibrium** $0 = \partial_{\mathbf{u}} \mathscr{E}_H(t_n, \mathbf{u}_H^n, \mathbf{v}_h^n, \mathbf{z}_h^n)$ reads in variational form

$$\int_{\Xi_H} \frac{1}{|\mathscr{Y}_\xi|} \int_{\mathscr{Y}_\xi} \partial_{\boldsymbol{\varepsilon}} W(\boldsymbol{\varepsilon}_{\xi,H}^n + \boldsymbol{\varepsilon}(\mathbf{v}_{\xi,h}^n), \mathbf{z}_{\xi,h}^n) : \boldsymbol{\varepsilon}(\delta\mathbf{u}_H) \, \mathrm{d}\mathbf{x} = \langle \ell^n, \delta\mathbf{u}_H \rangle$$

for $\delta\mathbf{u}_H \in V_H(\mathbf{0})$, where $\boldsymbol{\varepsilon}_{\xi,H}^n = \boldsymbol{\varepsilon}(\mathbf{u}_H^n)(\xi)$ denotes the macro-strain. We define the micro-stress

$$\sigma_{\xi,h}^n = \partial_{\boldsymbol{\varepsilon}} W(\boldsymbol{\varepsilon}_{\xi,H}^n + \boldsymbol{\varepsilon}(\mathbf{v}_{\xi,h}^n), \mathbf{z}_{\xi,h}^n)$$

depending on the macro-strain and the micro-fluctuation $\mathbf{v}_{\xi,h}^n \in V_{\xi,h}$ and then by averaging in the RVE the macro-stress

$$\sigma_{\xi,H}^n = \frac{1}{|\mathscr{Y}_\xi|} \int_{\mathscr{Y}_\xi} \sigma_{\xi,h}^n \, \mathrm{d}\mathbf{x} \,,$$

which together yields the macro-equilibrium in the form

$$\int_{\Xi_H} \sigma_{\xi,H}^n : \boldsymbol{\varepsilon}(\delta\mathbf{u}_H) = \langle \ell^n, \delta\mathbf{u}_H \rangle \,, \qquad \delta\mathbf{u}_H \in V_H(\mathbf{0}) \,.$$

The **Micro-Equilibrium** $0 = \partial_{\mathbf{v}} \mathscr{E}_H(t_n, \mathbf{u}_H^n, \mathbf{v}_h^n, \mathbf{z}_h^n)$ reads in variational form

$$\frac{1}{|\mathscr{Y}_\xi|} \int_{\mathscr{Y}_\xi} \sigma_{\xi,h}^n : \boldsymbol{\varepsilon}(\delta\mathbf{v}_{\xi,h}) \, \mathrm{d}\mathbf{x} = 0 \,, \qquad \delta\mathbf{v}_{\xi,h} \in V_{\xi,h} \,.$$

The **Flow Rule** $0 \in \partial_{\mathbf{z}} \mathscr{E}_H(t_n, \mathbf{u}_h^n, \mathbf{v}_h^n, \mathbf{z}_h^n) + \partial\mathscr{R}_H(\triangle\mathbf{z}_h^n)$ determines the history variable $\mathbf{z}_h^n$ from the macro-solution $\mathbf{u}_H^n$ and the micro-fluctuation $\mathbf{v}_{\xi,h}^n$. Depending on the conjugated variable $\mathbf{y}_{\xi,h}^n = -\partial_{\mathbf{z}} W(\boldsymbol{\varepsilon}_{\xi,H}^n + \boldsymbol{\varepsilon}(\mathbf{v}_{\xi,h}^n), \mathbf{z}_{\xi,h}^n)$ it is evaluated in every integration point of the RVE $\mathscr{Y}_\xi$ and can be expressed by duality

$$\mathbf{y}_{\xi,h}^n \in \partial\, R(\Delta \mathbf{z}_{\xi,h}^n) \qquad \Longleftrightarrow \qquad \Delta \mathbf{z}_{\xi,h}^n \in \partial\, R^*(\mathbf{y}_{\xi,h}^n) .$$

The two-scale damage problem We specify the incremental two-scale problem (4.17) for the damage model and we derive a generalised Newton method by inserting the results in Lemma 4.2. Here, for given damage history d_h^{n-1} and time t_n, the incremental two-scale problem aims to compute the unique minimiser $(\mathbf{u}_H^n, \mathbf{v}_h^n, d_h^n) \in V_H(\mathbf{u}_D^n) \times V_h \times Z_h$ of the functional

$$\mathscr{J}_{h,n}(\mathbf{u}_H^n, \mathbf{v}_h^n, d_h^n) = \mathscr{E}_H(t_n, \mathbf{u}_H^n, \mathbf{v}_h^n, d_h^n) + \mathscr{R}_H(d_h^n - d_h^{n-1})$$

by solving the nonlinear system

$$0 = \partial_{\mathbf{u}} \mathscr{E}_H(t_n, \mathbf{u}_H^n, \mathbf{v}_h^n, d_h^n) , \tag{4.18a}$$

$$0 = \partial_{\mathbf{v}} \mathscr{E}_H(t_n, \mathbf{u}_H^n, \mathbf{v}_h^n, d_h^n) , \tag{4.18b}$$

$$0 \in \partial_d \mathscr{E}_H(t_n, \mathbf{u}_H^n, \mathbf{v}_h^n, d_h^n) + \partial\, \mathscr{R}_H(\Delta d_h^n) . \tag{4.18c}$$

For this purpose, we define a return mapping procedure which evaluates the variational inequality (4.18c) which then allows to determine a suitable Newton linearisation.

The flow rule (4.18c) is evaluated at every integration point in the RVE $\mathscr{Y}_\xi$. Inserting $Y(\boldsymbol{\varepsilon}_{\xi,h}) = \tfrac{1}{2}\boldsymbol{\varepsilon}_{\xi,h} : \mathbb{C}[\boldsymbol{\varepsilon}_{\xi,h}]$, we obtain from Lemma 4.2 for the damage variable

$$d_{\xi,h,n}(\boldsymbol{\varepsilon}_{\xi,h}) = d_{\xi,h}^{n-1} + \max\left\{0,\, \Phi\big(Y(\boldsymbol{\varepsilon}_{\xi,h})\big) - d_{\xi,h}^{n-1}\right\}$$

and the stress response $\sigma_{\xi,h}^n = \sigma_{\xi,h,n}(\boldsymbol{\varepsilon}_{\xi,h}^n)$ with

$$\sigma_{\xi,h,n}(\boldsymbol{\varepsilon}_{\xi,h}) = \big(1 - d_{\xi,h,n}(\boldsymbol{\varepsilon}_{\xi,h})\big)\mathbb{C}[\boldsymbol{\varepsilon}_{\xi,h}] .$$

Inserting this result in (4.18) yields the reduced nonlinear problem to compute a critical point $(\mathbf{u}_H^n, \mathbf{v}_h^n) \in V_H(\mathbf{u}_D^n) \times V_h$ of

$$\langle \mathscr{F}_{h,n}(\mathbf{u}_H, \mathbf{v}_h),\, (\delta\mathbf{u}_H, \delta\mathbf{v}_h)\rangle =$$
$$\int_{\Xi_H} \frac{1}{|\mathscr{Y}_\xi|} \int_{\mathscr{Y}_\xi} \sigma_{\xi,h,n}(\boldsymbol{\varepsilon}_{\xi,h}) : \big(\boldsymbol{\varepsilon}(\delta\mathbf{u}_H) + \boldsymbol{\varepsilon}(\delta\mathbf{v}_{\xi,h})\big)\, d\mathbf{x} - \langle \ell^n, \delta\mathbf{u}_H\rangle$$

for all $(\delta\mathbf{u}_H, \delta\mathbf{v}_h) \in V_H(\mathbf{0}) \times V_h$. The consistent tangent operator

$$\mathbb{C}_{\xi,h,n}(\boldsymbol{\varepsilon}_{\xi,h}) = \big(1 - d_{\xi,h,n}(\boldsymbol{\varepsilon}_{\xi,h})\big)\mathbb{C}$$
$$- \operatorname{sgn}\left(\max\left\{0,\, \Phi(Y(\boldsymbol{\varepsilon}_{\xi,h})) - d_{\xi,h}^{n-1}\right\}\right)\Phi'(Y(\boldsymbol{\varepsilon}_{\xi,h}))\mathbb{C}[\boldsymbol{\varepsilon}_{\xi,h}] \otimes \mathbb{C}[\boldsymbol{\varepsilon}_{\xi,h}]$$

yields a Newton linearisation

$$\langle \mathscr{F}'_{h,n}(\mathbf{u}_H, \mathbf{v}_h)(\triangle\mathbf{u}_H, \triangle\mathbf{v}_h), (\delta\mathbf{u}_H, \delta\mathbf{v}_h)\rangle =$$

$$\int_{\Xi_H} \frac{1}{|\mathscr{Y}_\xi|} \int_{\mathscr{Y}_\xi} \mathbb{C}_{\xi,h,n}(\boldsymbol{\varepsilon}_{\xi,h})[\boldsymbol{\varepsilon}(\triangle\mathbf{u}_H) + \boldsymbol{\varepsilon}(\triangle\mathbf{v}_{\xi,h})] : (\boldsymbol{\varepsilon}(\delta\mathbf{u}_H) + \boldsymbol{\varepsilon}(\delta\mathbf{v}_{\xi,h})) \, d\mathbf{x}$$

for $(\triangle\mathbf{u}_H, \triangle\mathbf{v}_h), (\delta\mathbf{u}_H, \delta\mathbf{v}_h) \in V_H(\mathbf{0}) \times V_h$.

The two-scale residual $\mathscr{F}_{h,n}$ and its linearisation $\mathscr{F}'_{h,n}$ allows for the construction of a generalised Newton method of the incremental problem. This can be formulated as follows: starting with $(\mathbf{u}_H^{n,0}, \mathbf{v}_h^{n,0}) \in V_H(\mathbf{u}_D^n) \times V_h$, for $k = 1, 2, \dots$ the Newton increment $(\triangle\mathbf{u}_H^{n,k}, \triangle\mathbf{v}_h^{n,k}) \in V_H(\mathbf{0}) \times V_h$ is determined by solving

$$\langle \mathscr{F}'_{h,n}(\mathbf{u}_H^{n,k-1}, \mathbf{v}_h^{n,k-1})(\triangle\mathbf{u}_H^{n,k}, \triangle\mathbf{v}_h^{n,k}), (\delta\mathbf{u}_H, \delta\mathbf{v}_h)\rangle =$$
$$- \langle \mathscr{F}_{h,n}(\mathbf{u}_H^{n,k-1}, \mathbf{v}_h^{n,k-1}), (\delta\mathbf{u}_H, \delta\mathbf{v}_h)\rangle$$

for all $(\delta\mathbf{u}_H, \delta\mathbf{v}_h) \in V_H(\mathbf{0}) \times V_h$. The next iterate is given by

$$(\mathbf{u}_H^{n,k}, \mathbf{v}_h^{n,k}) = (\mathbf{u}_H^{n,k-1}, \mathbf{v}_h^{n,k-1}) + s_{n,k}(\triangle\mathbf{u}_H^{n,k}, \triangle\mathbf{v}_h^{n,k})$$

with a suitable damping parameter $s_{n,k} \in (0, 1]$. The iteration stops if the residual $\mathscr{F}_{h,n}(\mathbf{u}_H^{n,k}, \mathbf{v}_h^{n,k})$ is small enough.

It turns out that this monolithic Newton method for the combined two-scale problem is not efficient, since for every Newton step a full micro-macro problem has to be solved. So we use an alternative approach to compute the Newton increment first on the micro-scale and then on the macro-scale.

In the first time step $n = 0$, we compute for every $\xi \in \Xi_H$ the micro-fluctuations $\mathbf{w}_{\xi,h,1}^0, \dots, \mathbf{w}_{\xi,h,6}^0 \in V_{\xi,h}$ with respect to the basis $\eta_1, \dots, \eta_6$ solving (4.4).

In every time step $n \geq 1$ we set $\mathbf{w}_{\xi,h,l}^{n,0} = \mathbf{w}_{\xi,h,l}^{n-1}$ and we start with selecting $\mathbf{u}_H^{n,0} \in V_H(\mathbf{u}_D^n)$. For every macro-Newton step $k \geq 1$ and for every $\xi \in \Xi_H$, the micro-residual at $\boldsymbol{\varepsilon}_{\xi,H}^{n,k-1} = \boldsymbol{\varepsilon}(\mathbf{u}_H^{n,k-1})(\xi)$ is given by

$$\langle \mathscr{F}_{\xi,h,n,k}(\mathbf{v}_{\xi,h}), \delta\mathbf{v}_{\xi,h}\rangle = \int_{\mathscr{Y}_\xi} \sigma_{\xi,h,n}(\boldsymbol{\varepsilon}_{\xi,H}^{n,k-1} + \boldsymbol{\varepsilon}(\mathbf{v}_{\xi,h})) : \boldsymbol{\varepsilon}(\delta\mathbf{v}_{\xi,h}) \, d\mathbf{x} .$$

The micro-fluctuation $\mathbf{v}_{\xi,h}^{n,k}$ is computed by a micro-Newton method solving approximately the nonlinear problem $\mathscr{F}_{\xi,h,n,k}(\mathbf{v}_{\xi,h}) = 0$. Starting with

$$\mathbf{v}_{\xi,h}^{n,k,0} = \sum_{j=1}^{6} \left(\boldsymbol{\varepsilon}_{\xi,H}^{n,k-1} : \eta_j\right)\mathbf{w}_{\xi,h,j}^{n,k-1} ,$$

we compute for $m = 1, 2, \ldots$ the strain $\boldsymbol{\varepsilon}_{\xi,h}^{n,k,m-1} = \boldsymbol{\varepsilon}_{\xi,H}^{n,k-1} + \boldsymbol{\varepsilon}(\mathbf{v}_{\xi,h}^{n,k,m-1})$, the stress response $\sigma_{\xi,h}^{n,k,m-1} = \sigma_{\xi,h,n}(\boldsymbol{\varepsilon}_{\xi,h}^{n,k,m-1})$ and the consistent tangent operator $\mathbb{C}_{\xi,h}^{n,k,m-1} = \mathbb{C}_{\xi,h,n}(\boldsymbol{\varepsilon}_{\xi,h}^{n,k,m-1})$.

Then, the increment $\triangle\mathbf{v}_{\xi,h}^{n,k,m} \in V_{\xi,h}$ is computed by solving

$$\int_{\mathscr{Y}_\xi} \mathbb{C}_{\xi,h}^{n,k,m-1}[\boldsymbol{\varepsilon}(\triangle\mathbf{v}_{\xi,h}^{n,k,m})] : \boldsymbol{\varepsilon}(\delta\mathbf{v}_{\xi,h})\,d\mathbf{x} = -\int_{\mathscr{Y}_\xi} \sigma_{\xi,h}^{n,k,m-1} : \boldsymbol{\varepsilon}(\delta\mathbf{v}_{\xi,h})\,d\mathbf{x}$$

for all $\delta\mathbf{v}_{\xi,h} \in V_{\xi,h}$ defining $\mathbf{v}_h^{n,k,m} = \mathbf{v}_h^{n,k,m-1} + s_{\xi,n,k,m}\triangle\mathbf{v}_h^{n,k,m}$ with $s_{\xi,n,k,m} \in (0, 1]$. If the micro-residual is small enough, we set $\mathbf{v}_{\xi,h}^{n,k} = \mathbf{v}_{\xi,h}^{n,k,m}$, $\sigma_{\xi,h}^{n,k} = \sigma_{\xi,h}^{n,k,m}$,

$$\sigma_{\xi,H}^{n,k} = \frac{1}{|\mathscr{Y}_\xi|} \int_{\mathscr{Y}_\xi} \sigma_{\xi,h}^{n,k}\,d\mathbf{x},$$

and $\mathbb{C}_{\xi,h}^{n,k} = \mathbb{C}_{\xi,h}^{n,k,m}$. If $\mathbf{v}_{\xi,h}^{n,k}$ is sufficiently close to the previous iterate, we use $\mathbf{w}_{\xi,h,l}^{n,k} = \mathbf{w}_{\xi,h,l}^{n,k-1}$ and $\mathbb{C}_{\xi,H}^{n,k} = \mathbb{C}_{\xi,H}^{n,k-1}$ from the previous iteration, otherwise we compute $\mathbf{w}_{\xi,h,l}^{n,k} \in V_{\xi,h}$ solving

$$\frac{1}{|\mathscr{Y}_\xi|} \int_{\mathscr{Y}_\xi} \mathbb{C}(\mathbf{x})[\eta_l + \boldsymbol{\varepsilon}(\mathbf{w}_{\xi,h,l}^{n,k})] : \boldsymbol{\varepsilon}(\delta\mathbf{v}_{\xi,h})\,d\mathbf{x} = 0, \qquad \delta\mathbf{v}_{\xi,h} \in V_{\xi,h}$$

for $l = 1, \ldots, 6$ and the multi-scale tensor

$$\mathbb{C}_{\xi,H}^{n,k} = \frac{1}{|\mathscr{Y}_\xi|} \sum_{l,j=1}^{6} \left(\int_{\mathscr{Y}_\xi} \mathbb{C}_{\xi,h}^{n,k}[\eta_l + \boldsymbol{\varepsilon}(\mathbf{w}_{\xi,h,l}^{n,k})] : \eta_j\,d\mathbf{x} \right) \eta_l \otimes \eta_j.$$

The macro-update $\triangle\mathbf{u}_H^{n,k} \in V_H(\mathbf{0})$ is computed solving

$$\int_{\varXi_H} \mathbb{C}_{\xi,H}^{n,k}[\boldsymbol{\varepsilon}(\triangle\mathbf{u}_H^{n,k})] : \boldsymbol{\varepsilon}(\delta\mathbf{u}_H) = -\int_{\varXi_H} \sigma_{\xi,H}^{n,k} : \boldsymbol{\varepsilon}(\delta\mathbf{u}_H) + \langle \ell^n, \delta\mathbf{u}_H \rangle$$

for all $\delta\mathbf{u}_H \in V_H(\mathbf{0})$ defining $\mathbf{u}_H^{n,k} = \mathbf{u}_H^{n,k-1} + s_{n,k}\triangle\mathbf{u}_h^{n,k}$ with $s_{n,k} \in (0, 1]$. If the macro-residual is small enough, we set $\mathbf{u}_H^n = \mathbf{u}_H^{n,k}$, $\mathbf{v}_{\xi,h}^n = \mathbf{v}_{\xi,h}^{n,k}$, we update the damage variable $d_{\xi,h}^n = d_{\xi,h}^{n-1} + \max\left\{0, \Phi\left(Y_\xi(\boldsymbol{\varepsilon}_{\xi,h}^n)\right) - d_{\xi,h}^{n-1}\right\}$, and then we proceed to the next time step.

The two-scale elasto-plasticity model Specifying the incremental two-scale problem (4.17) for elasto-plasticity yields

$$0 = \partial_\mathbf{u}\mathscr{E}(t_n, \mathbf{u}_H^n, \mathbf{v}_h^n, \boldsymbol{\varepsilon}_{p,h}^n, r_h^n), \tag{4.19a}$$

$$0 = \partial_\mathbf{v}\mathscr{E}(t_n, \mathbf{u}_H^n, \mathbf{v}_h^n, \boldsymbol{\varepsilon}_{p,h}^n, r_h^n), \tag{4.19b}$$

$$0 \in \partial_{(\boldsymbol{\varepsilon}_p,r)}\mathscr{E}(t_n, \mathbf{u}_H^n, \mathbf{v}_h^n, \boldsymbol{\varepsilon}_{p,h}^n, r_h^n) + \partial\mathscr{R}(\triangle\boldsymbol{\varepsilon}_{p,h}^n, \triangle r_h^n). \tag{4.19c}$$

For given material history $(\boldsymbol{\varepsilon}_{p,\xi,h}^{n-1}, r_{\xi,h}^{n-1})$ and strain $\boldsymbol{\varepsilon}_{\xi,h}^n = \boldsymbol{\varepsilon}(\mathbf{u}_H^n)(\xi) + \boldsymbol{\varepsilon}(\mathbf{v}_{\xi,h}^n)$, the stress $\sigma_{\xi,h}^n = \sigma_{\xi,h,n}(\boldsymbol{\varepsilon}_{\xi,h}^n)$ is determined from the flow rule, cf. Lemma 4.3. Therefore, we define the relative trial stress

$$\alpha_{\xi,h,n}^{\text{tr}}(\boldsymbol{\varepsilon}_{\xi,h}) = 2\mu \text{ dev } \boldsymbol{\varepsilon}_{\xi,h} - (2\mu + K)\boldsymbol{\varepsilon}_{p,\xi,h}^{n-1},$$

and the flow function

$$F_{\xi,h,n}(\Delta r_{\xi,h}; \boldsymbol{\varepsilon}_{\xi,h}) = |\alpha_{\xi,h,n}^{\text{tr}}(\boldsymbol{\varepsilon}_{\xi,h})| - (2\mu + K)\Delta r_{\xi,h} - \Psi(r_{\xi,h}^{n-1} + \Delta r_{\xi,h}).$$

If $F_{\xi,h,n}(0; \boldsymbol{\varepsilon}_{\xi,h}) \leq 0$, we set $\Delta r_{\xi,h,n}(\boldsymbol{\varepsilon}_{\xi,h}) = 0$, otherwise the increment is defined by solving the nonlinear problem $F_{\xi,h,n}(\Delta r_{\xi,h,n}(\boldsymbol{\varepsilon}_{\xi,h}); \boldsymbol{\varepsilon}_{\xi,h}) = 0$. This defines the update $r_{\xi,h,n}(\boldsymbol{\varepsilon}_{\xi,h}) = r_{\xi,h}^{n-1} + \Delta r_{\xi,h,n}(\boldsymbol{\varepsilon}_{\xi,h})$ and the response for the plastic strain and the stress, and the consistent tangent operator

$$\boldsymbol{\varepsilon}_{p,\xi,h}(\boldsymbol{\varepsilon}_{\xi,h}) = \boldsymbol{\varepsilon}_{p,\xi,h}^{n-1} + \Delta r_{\xi,h,n}(\boldsymbol{\varepsilon}_{\xi,h}) \frac{\alpha_{\xi,h,n}^{\text{tr}}(\boldsymbol{\varepsilon}_{\xi,h})}{|\alpha_{\xi,h,n}^{\text{tr}}(\boldsymbol{\varepsilon}_{\xi,h})|},$$

$$\sigma_{\xi,h,n}(\boldsymbol{\varepsilon}_{\xi,h}) = \mathbb{C}[\boldsymbol{\varepsilon}_{\xi,h} - \boldsymbol{\varepsilon}_{p,\xi,h}(\boldsymbol{\varepsilon}_{\xi,h})],$$

$$\mathbb{C}_{\xi,h,n}(\boldsymbol{\varepsilon}_{\xi,h}) = \mathbb{C} - \frac{4\mu^2 \Delta r_{\xi,h,n}(\boldsymbol{\varepsilon}_{\xi,h})}{|\alpha_{\xi,h,n}(\boldsymbol{\varepsilon}_{\xi,h})|}\left(\text{dev} - \frac{\alpha_{\xi,h,n}(\boldsymbol{\varepsilon}_{\xi,h})}{|\alpha_{\xi,h,n}(\boldsymbol{\varepsilon}_{\xi,h})|} \otimes \frac{\alpha_{\xi,h,n}(\boldsymbol{\varepsilon}_{\xi,h})}{|\alpha_{\xi,h,n}(\boldsymbol{\varepsilon}_{\xi,h})|}\right)$$
$$- \frac{4\mu^2}{2\mu + K + \Psi'(r_{\xi,h,n}(\boldsymbol{\varepsilon}_{\xi,h}))} \frac{\alpha_{\xi,h,n}(\boldsymbol{\varepsilon}_{\xi,h})}{|\alpha_{\xi,h,n}(\boldsymbol{\varepsilon}_{\xi,h})|} \otimes \frac{\alpha_{\xi,h,n}(\boldsymbol{\varepsilon}_{\xi,h})}{|\alpha_{\xi,h,n}(\boldsymbol{\varepsilon}_{\xi,h})|}.$$

Now, the residual, $\mathscr{F}_{n,h}$, the linearisation $\mathscr{F}_{n,h}'$, and the generalised Newton method for the macro-problem can be defined as for the damage model, and inserting the elasto-plastic stress response yields the corresponding two-scale system for the combined model.

The two-scale model combining damage and elasto-plasticity Here, the internal variable has $N = 7$ components $\mathbf{z}_{\xi,h}^n = (d_{\xi,h}^n, \boldsymbol{\varepsilon}_{p,\xi,h}^n, r_{\xi,h}^n)$, and in (4.19a) we use the combined energy and dissipation functional, cf. Sect. 4.4. Inserting the stress response from Lemma 4.4 and the corresponding consistent tangent yields the residual and its linearisation as in the previous cases.

Parallel nonlinear two-scale algorithms The full algorithm is realised in three loops (see Fig. 4.4 for an overview): the outer loop for the time stepping, the Newton iteration for the macro-problem for every incremental problem, and in the inner loop the Newton iterations for the micro-problem for every RVE evaluating the local stress response.

We extend the parallel algorithm in Sect. 4.2 to inelastic applications. Every Newton iteration in the incremental problem has the structure of the linear two-scale model, provided that the residual and the consistent tangent is evaluated in every RVE. To obtain an efficient method, we use heuristic criteria in the RVE whether a new multi-scale basis is required.

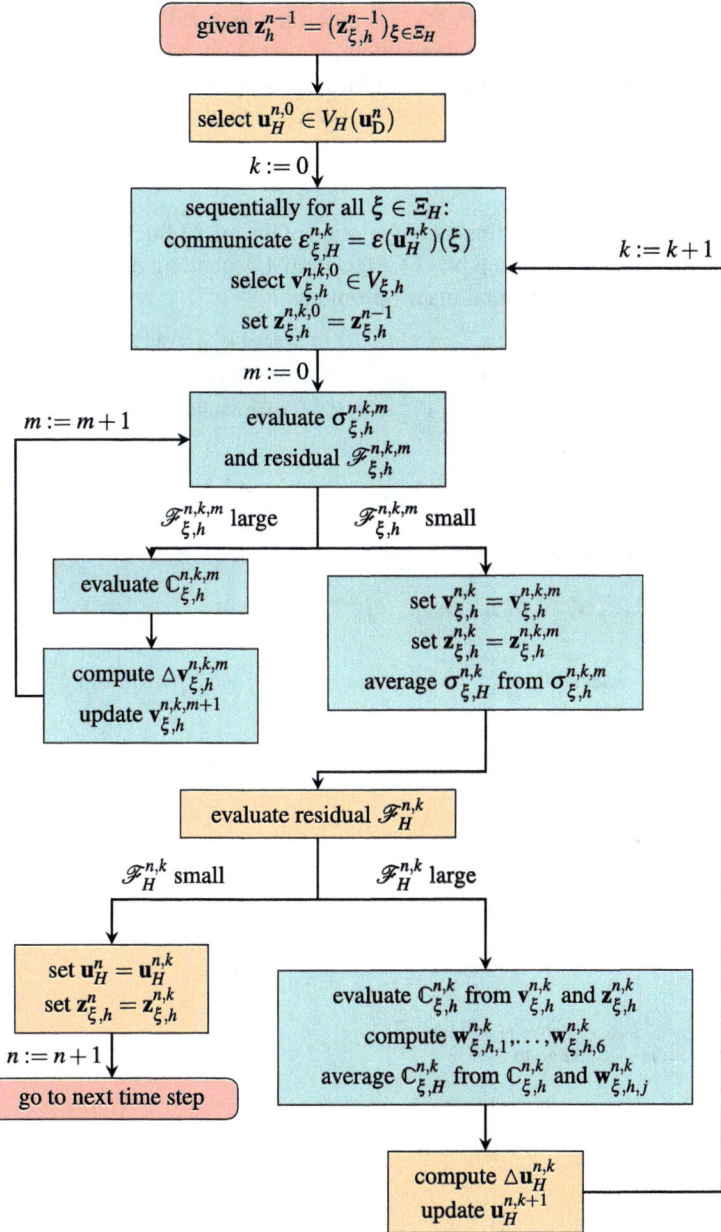

Fig. 4.4 The parallel incremental two-scale algorithm

The Newton method in every RVE is described in detail in Algorithm 3. This is used in the full parallel two-scale method in Algorithm 4 for the evaluation of the residual and for the computation of the linearisation on macro-scale. The damping factors in S5) and N7) are chosen by a line search strategy such that the next residual is decreasing.

Algorithm 3: Nonlinear computation of the micro-fluctuation $\mathbf{v}_{\xi,h}^{n,k}$ at time t_n and Newton iteration k in step N1) of Algorithm 4 depending on the strain approximation $\boldsymbol{\varepsilon}_{\xi,H}^{n,k}$ of the incremental macro-problem.

S0) On p with $\xi \in \Omega^p$ evaluate $\boldsymbol{\varepsilon}_{\xi,H}^{n,k} = \boldsymbol{\varepsilon}(\mathbf{u}_H^{n,k})(\xi)$ and send $\boldsymbol{\varepsilon}_{\xi,H}^{n,k}$ to all processes.
   Set $\mathbf{v}_{\xi,h}^{n,k,0} = \sum_l \left(\boldsymbol{\varepsilon}_{\xi,H}^{n,k} : \eta_l \right) \mathbf{w}_{\xi,h,l}^{n,k-1}$, $\mathbb{C}_{\xi,h}^{n,k,0} = \mathbb{C}_{\xi,h}^{n,k}$, and $m = 0$.

S1) Evaluate the micro-strain $\boldsymbol{\varepsilon}_{\xi,h}^{n,k,m} = \boldsymbol{\varepsilon}_{\xi,H}^{n,k} + \boldsymbol{\varepsilon}(\mathbf{v}_{\xi,h}^{n,k,m})$, the nonlinear material response $\mathbf{z}_{\xi,h}^{n,k,m} = \mathbf{z}_{\xi,h,n}(\boldsymbol{\varepsilon}_{\xi,h}^{n,k,m})$, the micro-stress

$$\sigma_{\xi,h}^{n,k,m} = \partial_{\boldsymbol{\varepsilon}} W(\boldsymbol{\varepsilon}_{\xi,h}^{n,k,m}, \mathbf{z}_{\xi,h}^{n,k,m}),$$

and the micro-residual

$$\langle \mathscr{F}_h(\mathbf{v}_{\xi,h}^{n,k,m}), \delta \mathbf{v}_{\xi,h} \rangle = \int_{\mathscr{Y}_{\xi}} \sigma_{\xi,h}^{n,k,m} : \boldsymbol{\varepsilon}(\delta \mathbf{v}_{\xi,h}) \, d\mathbf{x}, \qquad \delta \mathbf{v}_{\xi,h} \in V_{\xi,h}.$$

S2) If the micro-residual $\mathscr{F}_h(\mathbf{v}_{\xi,h}^{n,k,m})$ is small enough, set $\mathbf{v}_{\xi,h}^{n,k} = \mathbf{v}_{\xi,h}^{n,k,m}$, $\sigma_{\xi,h}^{n,k} = \sigma_{\xi,h}^{n,k,m}$ and $\mathbb{C}_{\xi,h}^{n,k} = \mathbb{C}_{\xi,h}^{n,k,m}$, and go to N2).

S3) If $m = m_{\max}$, reduce $\triangle t_n$ and go to T1).

S4) Evaluate the consistent tangent operator $\mathbb{C}_{\xi,h}^{n,k,m} = \mathbb{C}_{\xi,h,n}(\boldsymbol{\varepsilon}_{\xi,h}^{n,k,m})$ and compute $\triangle \mathbf{v}_{\xi,h}^{n,k,m} \in V_{\xi,h}$ solving in parallel

$$\int_{\mathscr{Y}_{\xi}} \mathbb{C}_{\xi,h}^{n,k,m}[\boldsymbol{\varepsilon}(\triangle \mathbf{v}_{\xi,h}^{n,k,m})] : \boldsymbol{\varepsilon}(\delta \mathbf{v}_{\xi,h}) \, d\mathbf{x} = -\langle \mathscr{F}_h(\mathbf{v}_{\xi,h}^{n,k,m}), \delta \mathbf{v}_{\xi,h} \rangle, \qquad \delta \mathbf{v}_{\xi,h} \in V_{\xi,h}.$$

S5) Select a damping parameter $s_{n,k,m} \in (0, 1]$ and set

$$\mathbf{v}_{\xi,h}^{n,k,m+1} = \mathbf{v}_{\xi,h}^{n,k,m} + s_{n,k,m} \triangle \mathbf{v}_{\xi,h}^{n,k,m}.$$

   If $s_{n,k,m} \leq s_{\min}$, reduce $\triangle t_n$ and go to T1).

S6) Set $m := m + 1$ and go to S1).

Algorithm 4: Parallel heterogeneous two-scale method for inelastic materials using Algorithm 3 in $\mathscr{Y}_\xi$.

T0) For all points $\xi \in \Xi_H$ with representative micro-structure compute in parallel the micro-fluctuations $\mathbf{w}^0_{\xi,h,1}, \ldots, \mathbf{w}^0_{\xi,h,6} \in V_{\xi,h}$ solving

$$\int_{\mathscr{Y}_\xi} \mathbb{C}(\mathbf{x})[\eta_l + \boldsymbol{\varepsilon}(\mathbf{w}^0_{\xi,h,l})] : \boldsymbol{\varepsilon}(\delta\mathbf{v}_{\xi,h}) \, d\mathbf{x} = 0, \qquad \delta\mathbf{v}_{\xi,h} \in V_{\xi,h},$$

and compute the elastic multi-scale tensor

$$\mathbb{C}^0_{\xi,H} = \frac{1}{|\mathscr{Y}_\xi|} \sum_{l,j=1}^6 \left(\int_{\mathscr{Y}_\xi} \mathbb{C}(\mathbf{x})[\eta_l + \boldsymbol{\varepsilon}(\mathbf{w}^0_{\xi,h,l})] : \eta_j \, d\mathbf{x} \right) \eta_l \otimes \eta_j.$$

Set $\mathbf{z}^0_h = \mathbf{0}$, $t_0 = 0$, and $n = 1$.

T1) For given history variable $\mathbf{z}^{n-1}_h$ and time increment $\Delta t_n \in (0, T - t_{n-1})$ set $t_n = t_{n-1} + \Delta t_n$ and compute the following steps:

N0) Set $\mathbf{u}^{n,0}_H = \mathbf{u}^{n-1}_H$, $\mathbf{z}^{n,0}_{\xi,h} = \mathbf{z}^{n-1}_{\xi,h}$, $\mathbb{C}^{n,0}_{\xi,h} = \mathbb{C}^{n-1}_{\xi,h}$, $\mathbf{w}^{n,0}_{\xi,h,l} = \mathbf{w}^{n-1}_{\xi,h,l}$, and $k = 0$.
Set Dirichlet data $\mathbf{u}^{n,0}_H(\mathbf{x}) = \mathbf{u}_D(\mathbf{x}, t_n)$ on all nodal points $\mathbf{x} \in \partial\Omega_D$ of the macro-space V_H.

N1) Evaluate the macro-strain $\boldsymbol{\varepsilon}^{n,k}_H = \boldsymbol{\varepsilon}(\mathbf{u}^{n,k}_H)$ and compute the micro-fluctuation for all $\xi \in \Xi_H$ by Algorithm 3.

N2) Compute the macro-stress

$$\sigma^{n,k}_{\xi,H} = \frac{1}{|\mathscr{Y}_\xi|} \int_{\mathscr{Y}_\xi} \sigma^{n,k}_{\xi,h} \, d\mathbf{x}$$

and the macro-residual

$$\langle \mathscr{F}_H(\mathbf{u}^{n,k}_H), \delta\mathbf{u}_H \rangle = \int_{\Xi_H} \sigma^{n,k}_{\xi,H} : \boldsymbol{\varepsilon}(\delta\mathbf{u}_H) - \langle \ell^n, \delta\mathbf{u}_H \rangle, \qquad \delta\mathbf{u}_H \in V_H(\mathbf{0}).$$

N3) If macro-residual small enough, set $\mathbf{u}^n_H = \mathbf{u}^{n,k}_H$, $\mathbf{z}^n_h = \mathbf{z}^{n,k}_h$, $n := n+1$, and go to T1).

N4) If $k = k_{\max}$, reduce Δt_n and go to T1).

N5) Compute the micro-fluctuations $\mathbf{w}^{n,k}_{\xi,h,1}, \ldots \mathbf{w}^{n,k}_{\xi,h,6} \in V_{\xi,h}$ solving in parallel

$$\int_{\mathscr{Y}_\xi} \mathbb{C}^{n,k}_{\xi,h}[\eta_l + \boldsymbol{\varepsilon}(\mathbf{w}^{n,k}_{\xi,h,l})] : \boldsymbol{\varepsilon}(\delta\mathbf{v}_{\xi,h}) \, d\mathbf{x} = 0, \qquad \delta\mathbf{v}_{\xi,h} \in V_{\xi,h}.$$

Then compute the inelastic multi-scale tensor

$$\mathbb{C}^{n,k}_{\xi,H} = \sum_{l,j} \left(\frac{1}{|\mathscr{Y}_\xi|} \int_{\mathscr{Y}_\xi} \mathbb{C}^{n,k}_{\xi,h}[\eta_l + \boldsymbol{\varepsilon}(\mathbf{w}^{n,k}_{\xi,h,l})] : \eta_j \, d\mathbf{x} \right) \eta_l \otimes \eta_j.$$

N6) Compute $\Delta\mathbf{u}^{n,k}_H \in V_H(\mathbf{0})$ solving in parallel

$$\int_{\Xi_H} \mathbb{C}^{n,k}_{\xi,H}[\boldsymbol{\varepsilon}(\Delta\mathbf{u}^{n,k}_H)] : \boldsymbol{\varepsilon}(\delta\mathbf{u}_H) = -\langle \mathscr{F}_H(\mathbf{u}^{n,k}_H), \delta\mathbf{u}_H \rangle, \qquad \delta\mathbf{u}_H \in V_H(\mathbf{0}).$$

N7) Select a damping parameter $s_{n,k} \in (0, 1]$ and set

$$\mathbf{u}^{n,k+1}_H = \mathbf{u}^{n,k}_H + s_{n,k}\Delta\mathbf{u}^{n,k}_H.$$

If $s_{n,k} \leq s_{\min}$ reduce Δt_n and go to T1).

N8) Set $k := k+1$ and go to N1).

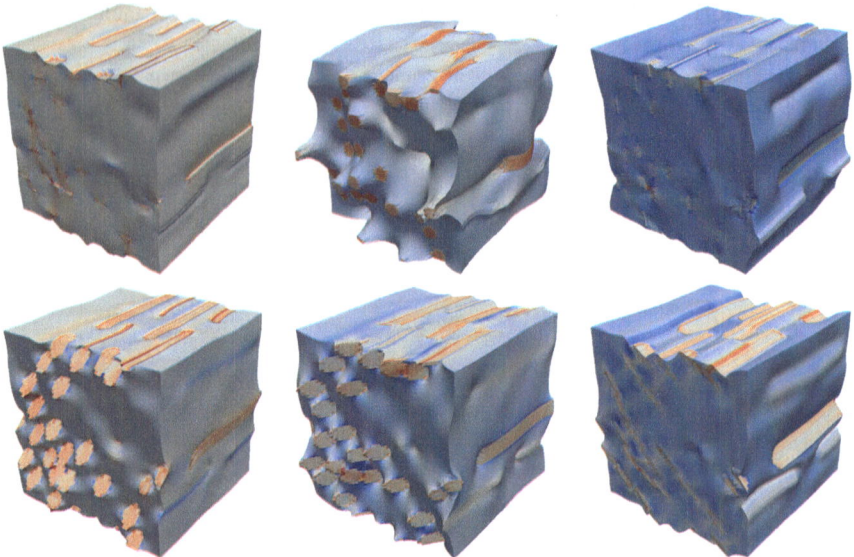

Fig. 4.5 Deformation of the periodic micro-fluctuations $\mathbf{w}^0_{\xi,h,k}$ corresponding to the basis tensors η_k in Sym(3) and stress distribution in the RVE

4.6 Numerical Experiments for Inelastic Material Models

The inelastic two-scale method is now applied to fibre reinforced polymers, again using the test configuration in Fig. 4.2 with boundary conditions (4.7). The material parameters for the inelastic models are taken from [27]. The damping and yielding point parameter in the damage model for the polymer is set to $H = 0.22702$ and $Y_0 = 0.08692$, and for isotropic plasticity we use yield strength $\sigma_y = 25$ and isotropic linear hardening law with parameters $H_0 = 1$ and $K_\infty - K_0 = 0$.

The two-scale damage model In the first experiment we investigate the inelastic uniaxial tensile test with the damage model, using a fibre reinforced micro-structure with 10% fibre volume fraction and a fibre orientation of 90°. We use 1024 integration points for the approximation of the macro-solution, and in every RVE a discretisation with dim $V_{\xi,h} = 823\,875$ for the representation of the micro-fluctuations $\mathbf{v}^n_{\xi,h}$ and dim $Z_{\xi,h} = 2\,097\,152$ to represent the variable $d^n_{\xi,h}$ at every integration point in the RVE.

At the beginning in step T0) of Algorithm 4, corresponding to the 6 symmetric tensor basis η_k, the representative micro-fluctuations $\mathbf{w}^0_{\xi,h,k}$ are computed determining the averaged linear material, see Fig. 4.5.

Starting with $d^0 = 0$, this is used to compute the elastic material response in every loading step for all RVE until the material response gets inelastic, see Fig. 4.6 for an example at a sample point.

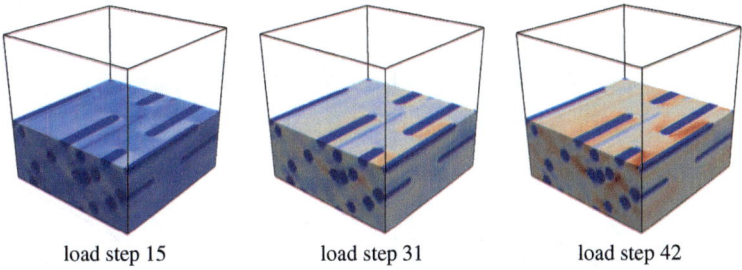

load step 15	load step 31	load step 42

Fig. 4.6 Evolution of the damage variable $d^n_{\xi,h}(\varepsilon^n_{\xi,h})$ in the RVE at the sample point $\xi = (0.115470, 0.302831, 1.183013)^\top$

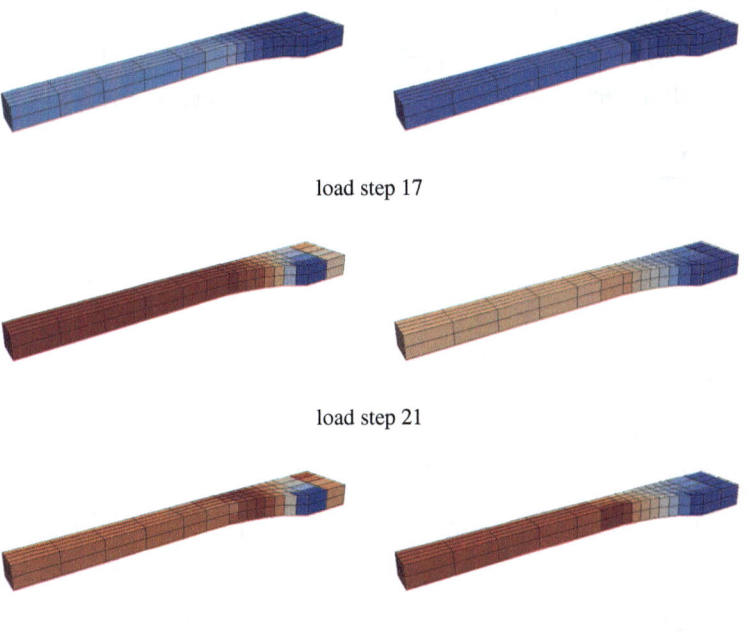

load step 17

load step 21

load step 43

Fig. 4.7 Evolution of the stress (left) and the damage variable (right) in the tensile test with 10% fibre volume fraction and fibre orientation of 90°

The macroscopic evolution of the averaged stress and damage variable

$$\sigma^n_{\xi,H} = \frac{1}{|\mathscr{Y}_\xi|} \int_{\mathscr{Y}_\xi} \sigma^n_{\xi,h} \, \mathrm{d}\mathbf{x} \,, \qquad d^n_{\xi,H} = \frac{1}{|\mathscr{Y}_\xi|} \int_{\mathscr{Y}_\xi} d^n_{\xi,h} \, \mathrm{d}\mathbf{x}$$

is shown in Fig. 4.7. Finally, the material response in all RVEs gets inelastic.

The overall simulation for $t \in [0, 132]$ requires 43 loading increments with 42 735 evaluations of the effective algorithmic tangent $\mathbb{C}^n_{\xi,H}$ with 216 894 Newton iterations

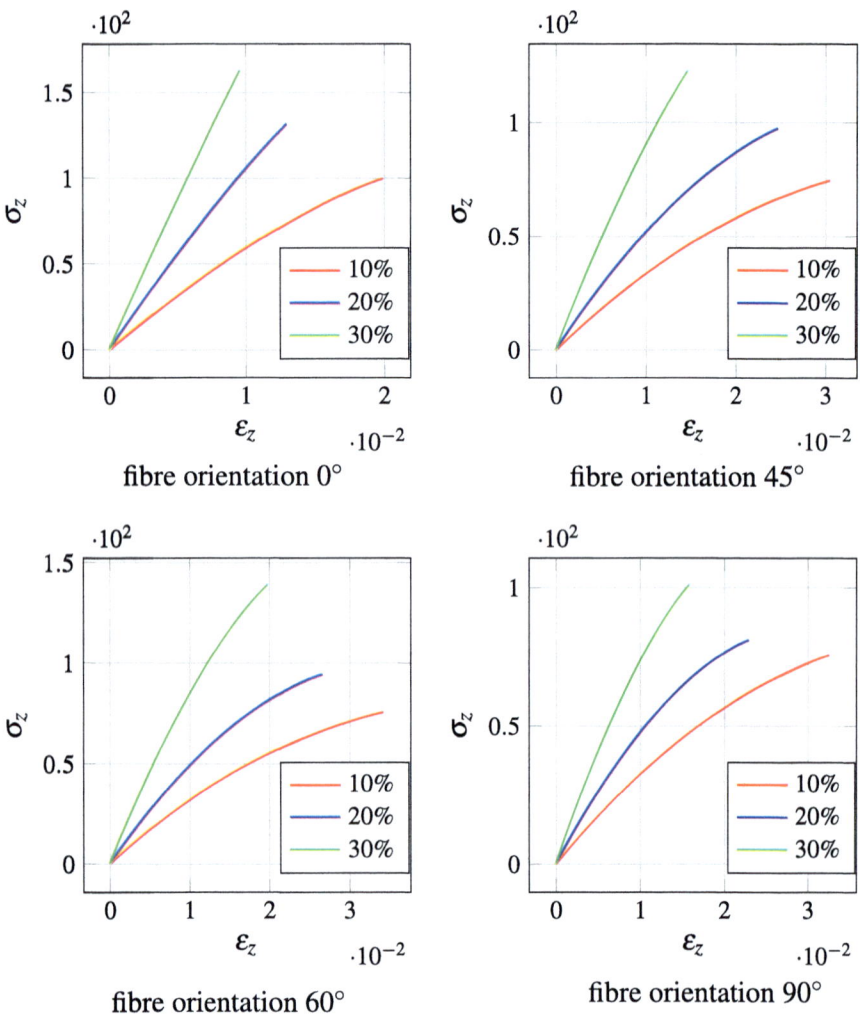

Fig. 4.8 Stress-strain curves of a uniaxial monotonic tensile test for a unidirectional short fibre reinforced material with different fibre volume fractions and fibre orientations. Here, stress σ_z and strain ε_z are evaluated by (4.8)

for the computation of the micro-fluctuation in Algorithm 3. The computation takes 4d 22h 41min on the ForHLR II cluster[4] with $512 = 32 \times 16$ cores.

Various fibre orientations and filler contents In the next test we investigate the inelastic material response of the damage model for different fibre orientations and volume fractions, see the stress-strain curves in Fig. 4.8. We clearly observe that the strength of the material is increased by a large volume fraction of the fibres, and

---

[4]https://www.scc.kit.edu/dienste/forhlr2.php.

the damage process is stronger for a fibre orientation orthogonal to the applied load. The simple damage model is limited to moderate loads, cf. Remark 4.4.1, so we stop the incremental test when the overall amount of damage is getting too large and the algorithmic tangent gets indefinite.

Comparison of inelastic two-scale models The simple damage model is not sufficient for a realistic description of fibre reinforced polymers. Experimental data in [21, Chap. 2.2, Fig. 6] exhibit in addition to the damage process characteristics of an elasto-plastic yield limit and hardening effects. For the numerical investigation of the different inelastic effects we consider cyclic loading using an uniaxial displacement driven load at $x_3 = 6.5$ with

$$
\mathbf{u}(t, \mathbf{x}) = \begin{cases} (t - T_{k-1})\mathbf{u}_0 & T_{k-1} < t < T_k \text{ loading,} \\ (T_k - t)\mathbf{u}_0 & T_k < t < T_{k+1} \text{ unloading,} \end{cases} \qquad \mathbf{u}_0 = u_0 \begin{pmatrix} 0 \\ 0 \\ 1 \end{pmatrix}
$$

for the transition points $T_0 = 0 < T_1 < T_2 < \cdots$ from loading to unloading and from unloading to loading. The scaling factor is set to $u_0 = 0.01$.

For the comparison of the different models we compute several load cycles, see Fig. 4.9. The transition points $T_1, T_3, T_5, \ldots$ are chosen such that the maximal stress is increased in every load cycle, and for complete unloading we set $T_{2k} = 2T_{2k-1} - T_{2k-2}$. We select an unidirectional micro-structure with 10% fibre volume fraction and orientation aligned to the traction force. In all cases we use on the macro-scale 128 sample points and in each RVE a discretisation with dim $V_{\xi,h} = 107\,811$ and 262 144 integration points for the representation of the memory variables $\mathbf{z}_{\xi,h}^n$.

For the inelastic evolution the simulation of the damage model requires 1 003 loading increments for 4 load cycles with together 457 012 Newton iterations for the computation of the micro-fluctuation, and 73 155 evaluations of the effective material response. Finally, the response in each sample point is inelastic. The load is increased in every load cycle, so that the stress response is more and more reduced, but no permanent deformation remains after unloading, see Fig. 4.9a.

In the elasto-plastic model we compute 298 loading increments for 2 load cycles requiring 218 098 Newton iterations for the computation of the micro-fluctuation $\mathbf{v}_{\xi,h}^n$ and 94 420 evaluations of the material response. Here, the response in 126 of 128 sample points behaves inelastic. Since the equivalent strain is monotone increasing in every inelastic increment, a residual stress remains after unloading, which is clearly observed in Fig. 4.9b. Also the yield stress and the linear hardening is characterised by the stress-strain curve: the effective yield stress is linearly increased in every loading cycle, and the elastic unloading is shifted in parallel by the equivalent strain increment.

The results for the combined model are computed with 620 loading increments in 347 296 Newton iterations with 51 691 evaluations for the effective material response, see Fig. 4.9c. Here, we observe both defect mechanisms, the shift of the residual stress after unloading caused by hardening effects, and the decreasing stiffness in the elastic unloading since in every load cycle the overall damage is increased.

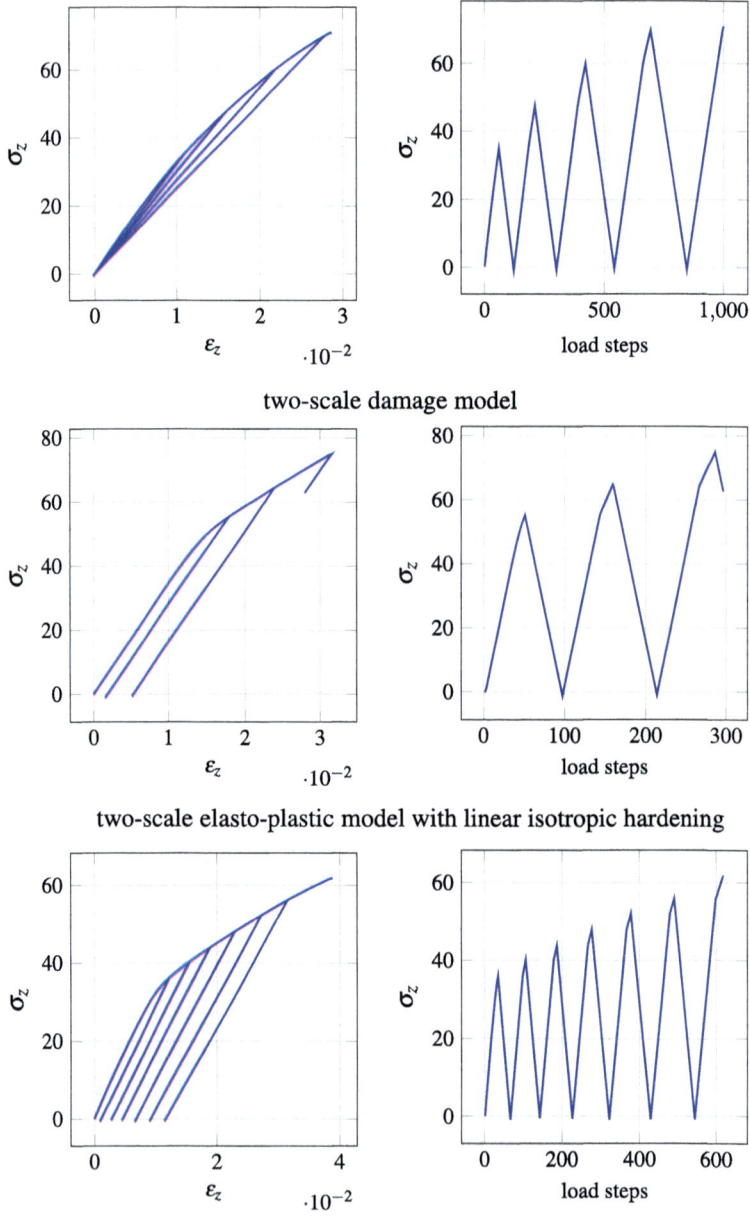

two-scale damage model

two-scale elasto-plastic model with linear isotropic hardening

two-scale model combining elasto-plasticity and damage

Fig. 4.9 Stress-strain curve (left) and stress-load step diagram (right) of uniaxial cyclic tensile tests with 10% fibre volume fraction and 90° fibre orientation with respect to the tensile load for a short fibre reinforced composite using different material models

Table 4.3 Numerical results for the macroscopic stress integral for the damage model with RVEs of different size

DoFs	σ_H^{δ}	$\sigma_H^{\delta/2}$	$\sigma_H^{\delta/4}$
375	95.739	73.746	63.559
2187	89.540	75.731	55.794
14739	85.206	64.988	50.780
107811	72.803	59.322	48.392
823875	66.679	56.745	47.373
6440067	63.955	55.558	46.654
$h \longrightarrow 0$	61.551	54.462	45.770

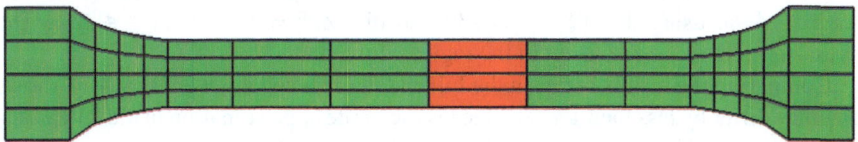

Fig. 4.10 Region of interest $\Omega_{\mathrm{ctr}} = (0, 0.5) \times (-0.2, 0.2) \times (0, 2) \subset \Omega$ with the full resolution of the RVEs

Comparing with experimental data [21, Chap. 2.2, Fig. 6] we observe that the two-scale model combining damage and plasticity is suitable for qualitatively correct description of the effective material behaviour, and that the coarse resolution on the macro-scale and the moderate resolution on the micro-scale is sufficient to capture these effects correctly.

Convergence test for the inelastic two-scale method In order to test whether the characteristic length scale resolution is sufficient, we compare the results of the tensile test for the damage model depending on the sample size δ and the mesh size h of the RVE, using an isotropic fibre distribution with 10% volume fraction, see Table 4.3. Comparing the results with the elastic case in Table 4.1 we observe that in the inelastic case the full resolution of the micro-structure is required, since for fractions of the RVE with $\delta/2$ and $\delta/4$ the material response is considerably different. Here, this is tested with fixed approximation on the macro-scale with dim $V_H = 165$, and for different δ the convergence with respect to the mesh size h on the micro-scale is considered. From this we can roughly estimate an accuracy of approximately 10% for the stress-strain curves in Fig. 4.9. For more details on the convergence analysis and the estimation of the extrapolated values for $h \longrightarrow 0$ we refer to [24, Chap. 7.3].

A reduced method The full simulation of the cyclic inelastic material behaviour is computationally very expensive. So we reduce the computational model by using a full resolution only in a region of interest $\Omega_{\mathrm{ctr}} \subset \Omega$ which is also used to evaluate the stress-strain curve by (4.8), see Fig. 4.10. Then, we use a fine resolution $V_{\xi,h}$ and $Z_{\xi,h}$ for $\mathscr{Y}_\xi \subset \Omega_{\mathrm{ctr}}$, and for $\mathscr{Y}_\xi \not\subset \Omega_{\mathrm{ctr}}$ coarser spaces $V_{\xi,h}^{\mathrm{red}}$ and $Z_{\xi,h}^{\mathrm{red}}$.

Fig. 4.11 Comparison of the cyclic stress-strain curves between the reduced (blue) and full (red) two-scale method

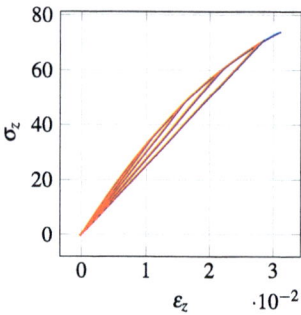

We evaluate the reduced model for the cyclic loading test with the damage model, see Fig. 4.9a, using dim $V_{\xi,h}^{red} = 14\,739$ and dim $Z_{\xi,h}^{red} = 14\,739$ for $\xi \notin \Omega_{ctr}$. The stress-strain curve is evaluated in the region of interest.

Comparing the results between the full and the reduced model in Fig. 4.11 shows that they differ by less then 1%, i.e., the less accurate approximation in the RVEs outside the region of interest has only a very small influence to the averaged macroscopic solution in Ω_{ctr}. The simulation for the reduced model with 938 time increments and together 439 814 Newton iterations in the RVEs requires 45 hours and for the full model approximately 3 days on the IC2 cluster[5] with with 256 cores distributed on 16 nodes.

Conclusion and outlook The full inelastic two-scale method of small-strain damage and plasticity models can be realised in an extremely large tensor product finite element space $V_H \times \prod_{\xi} V_{\xi,h}$, and in case of fine micro-structures small mesh sizes h are required. Since every RVE may have a different evolution, all memory variables $\prod_{\xi} Z_{\xi,h}$ need to be stored. Thus large parallel machines are required to realise this method and to represent the data well distributed. Here we propose a parallel solution scheme with an efficient parallel data representation and a stable two-stage nonlinear Newton method to determine the minimiser of the incremental loading step.

Nevertheless, on parallel machines with 1024 cores the full simulation of several loading cycles still requires a few days. For the next generation of high performance computers, our method has to be enhanced, see [3] for concepts to a flexible load balancing for a two-scale method applied to dual-phase steel. A further acceleration can be achieved by model reduction [6, 7], where the presented parallel two-scale method can be used in the offline phase in order to compute a suitable reduced basis. Our first test in Fig. 4.11 shows that is approach is promising.

---

[5]https://www.scc.kit.edu/dienste/ic2.php.

References

1. Allaire, G.: Homogenization and two-scale convergence. SIAM J. Math. Anal. **23**(6), 1482–1518 (1992)
2. Bakhvalov, N., Panasenko, G.: Homogenisation: Averaging Processes in Periodic Media : Mathematical Problems in the Mechanics of Composite Materials. Springer, Dordrecht (1989)
3. Balzani, D., Gandhi, A., Klawonn, A., Lanser, M., Rheinbach, O., Schröder, J.: One-way and fully-coupled FE2 methods for heterogeneous elasticity and plasticity problems: Parallel scalability and an application to thermo-elastoplasticity of dual-phase steels. In: Software for Exascale Computing-SPPEXA 2013–2015, pp. 91–112. Springer (2016)
4. Diebels, S., Jung, A., Chen, Z., Seibert, H., Scheffer, T.: Experimentelle Mechanik: Von der Messung zum Materialmodell. Rundbrief GAMM (2015)
5. Feyel, F., Chaboche, J.L.: FE2 multiscale approach for modelling the elastoviscoplastic behaviour of long fibre SiC/Ti composite materials. Comput. Methods Appl. Mech. Eng. **183**(3), 309–330 (2000)
6. Fritzen, F., Hodapp, M.: The finite element square reduced (FE2R) method with gpu acceleration: towards three-dimensional two-scale simulations. Int. J. Numer. Methods Eng. **107**(10), 853–881 (2016)
7. Fritzen, F., Hodapp, M., Leuschner, M.: GPU accelerated computational homogenization based on a variational approach in a reduced basis framework. Comput. Methods Appl. Mech. Eng. **278**, 186–217 (2014)
8. GeoDict: The digital material laboratory. http://www.geodict.de/ (2014)
9. Ju, J.: On energy-based coupled elastoplastic damage theories: constitutive modeling and computational aspects. Int. J. Solids Struct. **25**(7), 803–833 (1989)
10. Kachanov, L.: Introduction to Continuum Damage Mechanics. Springer, Mechanics of Elastic Stability (1986)
11. Lippmann, H., Lemaitre, J.: A Course on Damage Mechanics. Springer, Berlin Heidelberg (1996)
12. Maurer, D., Wieners, C.: A parallel block LU decomposition method for distributed finite element matrices. Parallel Comput. **37**(12), 742–758 (2011)
13. Miehe, C., Schotte, J., Schröder, J.: Computational micro-macro transitions and overall moduli in the analysis of polycrystals at large strains. Comput. Mater. Sci. **16**(1), 372–382 (1999)
14. Miehe, C., Schröder, J., Schotte, J.: Computational homogenization analysis in finite plasticity. Comput. Methods Appl. Mech. Eng. **171**, 3–4 (1999)
15. Mielke, A.: Evolution of rate-independent systems. In: Handbook of Differential Equations: Evolutionary Equations, vol. 2, chap. 6, pp. 461–559. North-Holland (2005)
16. Mielke, A., Roubíček, T.: Rate-Independent Systems: Theory and Application. Applied Mathematical Sciences. Springer, New York (2015)
17. Mielke, A., Timofte, A.M.: Two-scale homogenization for evolutionary variational inequalities via the energetic formulation. SIAM J. Math. Anal. **39**(2), 642–668 (2007)
18. Papanicolau, G., Bensoussan, A., Lions, J.: Asymptotic Analysis for Periodic Structures. Studies in Mathematics and its applications. Elsevier Science (1978)
19. Rabotnov, Y.: Creep Problems in Structural Members. Elsevier, Applied Mathematics and Mechanics Series (1969)
20. Röhrig, C.: Personal communication (2016)
21. Röhrig, C., Scheffer, T., Diebels, S.: Mechanical characterization of a short fiber-reinforced polymer at room temperature: experimental setups evaluated by an optical measurement system. In: Continuum Mechanics and Thermodynamics, pp. 1–19 (2017)
22. Sanchez-Palencia, E., Zaoui, A.: Homogenization techniques for composite media: lectures delivered at the CISM International Center for Mechanical Sciences, Udine, Italy, July 1–5, 1985. In: Lecture Notes in Physics. Springer (1987)
23. Schröder, J.: A numerical two-scale homogenization scheme: the FE2-method. In: Plasticity and Beyond: Microstructures. Crystal-Plasticity and Phase Transitions, pp. 1–64. Springer, Vienna (2014)

24. Shirazi Nejad, R.: A parallel elastic and inelastic heterogeneous multiscale method for rate-independent materials. Ph.D. thesis, Karlsruhe Institute of Technology (2017)
25. Simo, J., Hughes, T.: Computational Inelasticity. Interdisciplinary Applied Mathematics. Springer, New York (2000)
26. Smit, R., Brekelmans, W., Meijer, H.: Prediction of the mechanical behavior of nonlinear heterogeneous systems by multi-level finite element modeling. Comput. Methods Appl. Mech. Eng. **155**(1–2), 181–192 (1998)
27. Spahn, J.: An efficient multiscale method for modeling progressive damage in composite materials. Ph.D. thesis, Technische Universität Kaiserslautern (2015)
28. Spahn, J., Andrä, H., Kabel, M., Müller, R.: A multiscale approach for modeling progressive damage of composite materials using fast Fourier transforms. Comput. Methods Appl. Mech. Eng. **268**, 871–883 (2014)
29. Thomas, M., Mielke, A.: Damage of nonlinearly elastic materials at small strain—existence and regularity results. Zeitschrift Angewandte Mathematik und Mechanik **90**, 88–112 (2010)
30. Weinan, E., Engquist, B., et al.: The heterogeneous multiscale methods. Commun. Math. Sci. **1**(1), 87–132 (2003)
31. Wieners, C.: A geometric data structure for parallel finite elements and the application to multigrid methods with block smoothing. Comput. Vis. Sci. **13**(4), 161–175 (2010)

Chapter 5
Fast Boundary Element Methods
for Composite Materials

Richards Grzhibovskis, Christian Michel and Sergej Rjasanow

5.1 Introduction

Boundary element method (BEM) has become an indispensable tool for computing approximate solutions to many important boundary value problems (BVPs). The main advantages of this method are: the absence of a volume mesh, high accuracy on low order elements, numerical stability, and a low number of discrete unknowns. In its original form, BEM can be applied if the differential operator in question is linear, elliptic, and has constant coefficients. The BVPs arising from mechanics of composites often do not fit into these constraints. We consider a rather simple mechanical model involving material damage (see Sect. 5.2). It leads to a non-linear BVP with variable (even solution dependent) coefficients. Our approach to obtain a fast BEM for this BVP consists in the following steps. First, we reduce the non-linear BVP to a sequence of linear elliptic, non-homogeneous BVPs with constant coefficients. We then construct approximate particular solutions to the differential equations from the latter problems with the help of a new kind of radial basis functions (RBFs). These particular solutions allow us to reduce the non-homogeneous BVPs to homogeneous ones, which can be treated by the pure BEM.

This chapter is organised as follows. The formulation of the non-linear BVP, its reduction to a sequence of linear problems is described in Sect. 5.2. An approximate solution is then written with the help of discrete boundary integral operators while assuming that a particular solution is known. A way of constructing the necessary particular solution (or its approximation) is proposed in Sect. 5.3. An efficient numerical procedure for the solution of the RBF interpolation problem is described in Sect. 5.4. Theoretical results from Sects. 5.3 and 5.4 are illustrated with numerical examples in Sect. 5.5. Approximate solutions to the original nonlinear BVP are then constructed in the context of composite materials. The results are reported in Sect. 5.6.

R. Grzhibovskis · C. Michel · S. Rjasanow (✉)
University of Saarland, 66123 Saarbrücken, Germany
e-mail: rjasanow@num.uni-sb.de

© Springer-Verlag GmbH Germany, part of Springer Nature 2019
S. Diebels and S. Rjasanow (eds.), *Multi-scale Simulation of Composite Materials*,
Mathematical Engineering, https://doi.org/10.1007/978-3-662-57957-2_5

5.2   Mechanical Modelling

Consider a solid body modelled by a bounded Lipschitz domain $\Omega \subset \mathbb{R}^3$. Suppose, for simplicity, that a displacement field g_D is given on the boundary $\Gamma = \partial\Omega$. The displacement field $u \in (H^1(\Omega))^3$ inside the domain Ω is the solution of the Dirichlet BVP.[1]

$$\begin{cases} \operatorname{div} \sigma_d(u, x) = 0, & x \in \Omega, \\ \gamma_0 u(x) = g_D(x), & x \in \Gamma, \end{cases} \tag{5.1}$$

where

$$\sigma_d(u, x) = (1 - d(u, x, \text{hist}))\sigma(u, x)$$

is the stress field and

$$\gamma_0 : H^1(\Omega) \to H^{1/2}(\Gamma)$$

is the Dirichlet trace operator. The non-linear expression for the stress tensor σ_d was derived in [32]. The linear material response $\sigma(u, x) = \mathbf{C}\varepsilon(u, x)$ to the strain

$$\varepsilon(u, x) = \frac{1}{2}(\nabla u(x) + \nabla^\top u(x))$$

is reduced due to the displacement and history dependent damage $d(u, x, \text{hist})$ distributed inside the solid body. During an incremental application of the boundary conditions, this damage first appears at points, where the deformation energy

$$Y(u, x) = \varepsilon(u, x) : \mathbf{C}\varepsilon(u, x)$$

exceeds a certain material dependent threshold Y_0^2. Once it is occurred, the material remains damaged even if the load is reduced. The formula for the damage reads

$$d(u, x, \text{hist}) = \max_{\tau \in \text{hist}} (d(u, x, \tau), 1 - \exp(-H_0(\sqrt{Y(u, x)} - Y_0))),$$

where $d(u, x, \tau)$ denotes the damage during previous load increments. Taking a finite number of increments, we look for the equilibrium of forces while accounting for the history of the damage, i.e. we solve the following n_{inc} non-linear BVPs

$$\begin{cases} \operatorname{div} \sigma(u^i, x) = \operatorname{div}(d(u^i, x, (0, \tau_i))\sigma(u^i, x)), & x \in \Omega, \\ \gamma_0 u^i(x) = \tau_i g_D(x), & x \in \Gamma, \end{cases} \tag{5.2}$$

where $\tau_i = i/n_{inc}$ and $i = 1, \dots, n_{inc}$. To resolve the non-linearity, we construct a sequence $(u^{i,k})_{k \in \mathbb{N}}$ to approximate the function u^i for each load increment by setting

---

[1]One seeks for the solution u in the subspace $\{v \in (H^1(\Omega))^3 : \operatorname{div} \sigma_d(v, x) \in (L^2(\Omega))^3\}$ of the Sobolev space $(H^1(\Omega))^3$.

$u^{0,0} = 0$ and employing fixed point iterations

$$\begin{cases} \text{div } \sigma(u^{i,k+1}, x) = \text{div } (d(u^{i,k}, x, (0, \tau_i))\sigma(u^{i,k}, x)), \ x \in \Omega, \\ \gamma_0 u^{i,k+1}(x) = \tau_i g_D(x), \hspace{3.5cm} x \in \Gamma, \end{cases} \tag{5.3}$$

for $0 \le k \le K_i$ until the convergence criterion

$$\|u^{i,K_i+1} - u^{i,K_i}\|_{L^2(\Omega)} / \|u^{i,K_i+1}\|_{L^2(\Omega)} < \varepsilon_{\text{Stop}}$$

is fulfilled with some prescribed accuracy $\varepsilon_{\text{Stop}}$. The approximate solution for the load increment number i is then set as an initial approximation for the next increment, i.e. $u^{i+1,0} = u^{i,K_{i+1}}$.

In what follows, we describe a boundary integral approach for solving the linear problem (5.3) when the domain Ω either is isotropic and homogeneous (see Sect. 5.2.1) or consists of such parts (see Sect. 5.2.2). A new ansatz for approximating a particular solution is proposed later to avoid volume integrals. It is based on matrix valued radial basis functions, which are described in Sect. 5.3.

5.2.1 Single Domain Formulation

To perform a single fixed point iteration the linear Dirichlet BVP

$$\begin{cases} \text{div } \sigma(v, x) = f(x), \ x \in \Omega, \\ \gamma_0 v(x) = g(x), \ x \in \Gamma, \end{cases} \tag{5.4}$$

with $v = u^{i,k+1}$, $g = \tau_i g_D$, and $f(x) = \text{div } (d(u^{i,k}, x, (0, \tau_i))\sigma(u^{i,k}, x))$ must be solved. If the undamaged material is homogeneous and isotropic, it can be characterised by just two scalar parameters (Lamé constants) λ and μ. In this case the expression for stress reduces to

$$\sigma(v, x) = \lambda \text{ tr } \varepsilon(v, x)I + 2\mu \, \varepsilon(v, x),$$

and the equilibrium of forces is determined by the Lamé system

$$\mu \Delta v + (\lambda + \mu)\text{grad div } v = f. \tag{5.5}$$

Suppose a function v_P^f satisfies the Eq. (5.5). The solution to the BVP (5.4) can then be composed as the sum $v = v_B + v_P^f$, where the function v_B solves the homogeneous problem

$$\begin{cases} \mu \Delta v_B + (\lambda + \mu)\text{grad div } v_B = 0, \hspace{2cm} \text{in } \Omega, \\ \gamma_0 v_B(x) = g(x) - \gamma_0 v_P^f(x), \ x \in \Gamma. \end{cases} \tag{5.6}$$

Since the Lamé differential operator is elliptic with constant coefficients, its fundamental solution U^* is well known

$$U^*(x, y) = \frac{1 + \nu}{8\pi E(1 - \nu)} \left(\frac{3 - 4\nu}{|x - y|} I + \frac{(x - y)(x - y)^\top}{|x - y|^3} \right). \tag{5.7}$$

Here I is the three by three identity matrix. The Young modulus E and the Poisson's ratio ν are related to the Lamé constants through

$$\lambda = \frac{E\nu}{(1 + \nu)(1 - 2\nu)}, \quad \mu = \frac{E}{2(1 + \nu)}.$$

The solution to the problem (5.6) can be written with the help of the Neumann trace operator

$$\gamma_1 : H^1(\Omega) \to H^{-1/2}(\Gamma), \quad \gamma_1 w(x) = \lim_{\Omega \ni y \to x} \sigma(w, y)n(x) \tag{5.8}$$

and the boundary potentials

$$\tilde{V} : H^{-1/2}(\Gamma) \to H^1(\Omega), \quad (\tilde{V}w)(x) = \int_\Gamma U^*(x, y)w(y)\mathrm{d}S_y, \tag{5.9}$$

$$\tilde{K} : H^{1/2}(\Gamma) \to H^1(\Omega), \quad (\tilde{K}w)(x) = \int_\Gamma \gamma_{1,y} U^*(x, y)w(y)\mathrm{d}S_y. \tag{5.10}$$

It reads

$$v_B = \tilde{V}t - \tilde{K}(g - \gamma_0 v_P^f), \tag{5.11}$$

where the Neumann trace $t = \gamma_1 v_B$ represents traction forces on the boundary. This function is the unique solution to the boundary integral equation on Γ

$$(Vt) = \left(\frac{1}{2} + K \right) \left(g - \gamma_0 v_P^f \right), \tag{5.12}$$

with the single layer potential operator $V = \gamma_0 \tilde{V}$ and the double layer potential operator

$$K : H^{1/2}(\Gamma) \to H^{1/2}(\Gamma), \quad (Kw)(x) = \text{v.p.} \int_\Gamma \gamma_{1,y} U^*(x, y)w(y)\mathrm{d}S_y. \tag{5.13}$$

Thus, the missing ingredient in the representation formula (5.11) is expressed through the Dirichlet-to-Neumann mapping S (Steklov-Poincaré operator) as

$$t = S(g - \gamma_0 v_P^f),$$

where

$$S : H^{1/2}(\Gamma) \to H^{-1/2}(\Gamma), \quad S = V^{-1}\left(\frac{1}{2} + K\right). \tag{5.14}$$

This operator can also be written in the following symmetric form

$$S = D + \left(\frac{1}{2} + K'\right) V^{-1} \left(\frac{1}{2} + K\right), \tag{5.15}$$

with the help of the adjoint double layer potential operator K' and the hypersingular operator D, namely

$$K' : H^{1/2}(\Gamma) \to H^{1/2}(\Gamma), \quad (K'w)(x) = \text{v.p.} \int_{\Gamma} \gamma_{1,x} U^*(x, y)w(y)\mathrm{d}S_y, \tag{5.16}$$

$$D : H^{1/2}(\Gamma) \to H^{-1/2}(\Gamma), \quad Dw = -\gamma_1 \tilde{K}w. \tag{5.17}$$

We sum up the above computations and write the solution to the BVP (5.4) as

$$v = (\tilde{V}S - \tilde{K})(g - \gamma_0 v_P^f) + v_P^f. \tag{5.18}$$

We can now determine the reaction forces on the boundary Γ by computing the conormal derivative of the displacement v as

$$\bar{\gamma}_1 v(x) = \lim_{\Omega \ni y \to x} (1 - d(v, y, \text{hist}))\sigma(v, y)n(x) =$$

$$= (1 - \gamma_0 d(x))(S(g - \gamma_0 v_P^f) + \gamma_1 v_P^f), \tag{5.19}$$

where the material damage computed at the displacement field from the previous iteration is denoted by d, i.e.

$$d(x) = d(u^{i,k}, x, \text{hist}). \tag{5.20}$$

5.2.2 Multi-domain Formulation

We now consider a body with piecewise isotropic and homogeneous material properties. More precisely, we allow pairwise separated inclusions

$$\Omega_m, \ 1 \leq m \leq n_\Omega$$

to be present in the domain Ω, see Fig. 5.1. Their boundaries

$$\Gamma_m, \ 1 \leq m \leq n_\Omega$$

Fig. 5.1 Notations for the
multi domain formulation

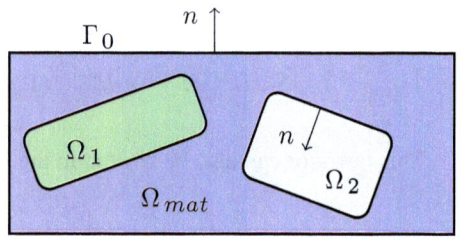

must not intersect the boundary $\Gamma_0 = \partial\Omega$. We denote by Ω_{mat} the domain occupied
by the matrix material, i.e.

$$\Omega_{mat} = \Omega \setminus \bigcup_{m=1}^{n_\Omega} \overline{\Omega}_m .$$

This implies the relation

$$\Gamma_{mat} = \partial\Omega_{mat} = \bigcup_{m=0}^{n_\Omega} \Gamma_m .$$

We now restate the BVP (5.3) in a form suitable for treatment with the boundary
integral method. As in the previous Subsection, denote the unknown function $u^{i,k+1}$
by v, the right hand side of the PDE by f, and the given boundary datum by g.

Due to the assumptions on the material properties, the displacement field v satisfies
the equation

$$\mu_m \Delta v + (\lambda_m + \mu_m)\text{grad div } v = f \quad \text{in } \Omega_m$$

for $m \in \{mat, 1, 2, \ldots, n_\Omega\}$ in each each subdomain and in the matrix material. Here
μ_m and λ_m are the Lamé constants of the corresponding material. The continuity
condition for the displacements on the internal boundaries reads

$$\gamma_0^{mat} v = \gamma_0^m v \text{ on } \Gamma_m , \ 1 \leq m \leq n_\Omega ,$$

where the upper index of the trace operator indicates the domain in which the limit
is taken i.e.

$$\gamma_0^m w(x) = \lim_{\Omega_m \ni y \to x} w(y), \quad x \in \Gamma_m .$$

Assuming that the damage occurs only in the matrix material, the equilibrium of
the traction forces on the internal boundaries can be expressed as

$$\bar{\gamma}_1^{mat} v = \gamma_1^m v \text{ on } \Gamma_m , \ 1 \leq m \leq n_\Omega ,$$

where

$$\bar{\gamma}_1^{mat} w(x) = (1 - \gamma_0^{mat} d(x))\gamma_1^{mat} w(x), \quad x \in \Gamma_{mat}$$

holds for any smooth w. The damage function in the above formula is determined by (5.20). Thus, the BVP for the case with isolated inclusions reads for $1 \leq m \leq n_\Omega$

$$
\begin{cases}
\mu_{mat} \Delta v + (\lambda_{mat} + \mu_{mat}) \text{grad div } v = f, & \text{in } \Omega_{mat}, \\
\mu_m \Delta v + (\lambda_m + \mu_m) \text{grad div } v = 0, & \text{in } \Omega_m, \\
\gamma_0^{mat} v(x) = \gamma_0^m v(x), \ x \in \Gamma_m, \\
\bar{\gamma}_1^{mat} v(x) = \gamma_1^m v(x), \ x \in \Gamma_m, \\
\gamma_0^{mat} v(x) = g(x), & x \in \Gamma_0.
\end{cases}
\tag{5.21}
$$

The second equation in the above system implies, that

$$
\gamma_1^m v = S_m \gamma_0^m v, \ m = 1, \ldots, n_\Omega.
$$

Here the Steklov-Poincaré operator S_m for the m-th inclusion is defined by (5.15) with the corresponding integration over the surface Γ_m. To find the displacement field in the matrix material, we first have to introduce the particular solution. Suppose, that we have constructed a function $v_P^f : \Omega_{mat} \to \mathbb{R}^3$ such that

$$
\mu_{mat} \Delta v_P^f + (\lambda_{mat} + \mu_{mat}) \text{grad div } v_P^f = f \quad \text{in } \Omega_{mat}
$$

holds. We now look for the unknown in the form $v = v_B + v_P^f$. The BVP (5.21) in these settings is reduced to the following $m + 1$ equations

$$
((1 - d) S_{mat} - S_m) \gamma_0^{mat} v_B = S_m \gamma_0^{mat} v_P^f - (1 - d) \gamma_1^{mat} v_P^f \quad \text{on } \Gamma_m,
\tag{5.22}
$$

$$
\gamma_0^{mat} v_B = g - \gamma_0^{mat} v_P^f \quad \text{on } \Gamma_0.
\tag{5.23}
$$

Because of the Dirichlet condition above, this system is uniquely solvable. The boundary Γ_{mat} is a union of mutually disjoint closed boundaries $\Gamma_m, 0 \leq m \leq n_\Omega$. Thus, we write the trace $\gamma_0^{mat} v$ as the sum

$$
\gamma_0^{mat} v(x) = \sum_{m=0}^{n_\Omega} v_m(x),
$$

where

$$
v_m(x) = \begin{cases} \gamma_0^{mat} v(x), \ x \in \Gamma_m, \\ 0, \ x \in \Gamma \setminus \Gamma_m. \end{cases}
$$

A substitution of this ansatz into (5.22) and the use of (5.23) yields the integral equations for all unknown functions v_m for $1 \leq m \leq n_\Omega$

$$
(1 - d) \sum_{j=1}^{n_\Omega} S_{mat} v_j - S_m v_m =
\tag{5.24}
$$

$$
S_m \gamma_0^{mat} v_P^f - (1 - d) \gamma_1^{mat} v_P^f + (1 - d)(S_{mat}(\gamma_0^{mat} v_P^f - g)) \quad \text{on } \Gamma_m.
$$

In the next subsection, we describe a symmetric Galerkin discretisation of the above equations on triangular meshes.

5.2.3 Galerkin Discretisation of the Integral Equations

We consider a sequence of boundary element meshes

$$\Gamma_N = \bigcup_{\ell=1}^{N} \overline{\tau}_\ell, \quad N \in \mathbb{N}. \tag{5.25}$$

In the most simple case, we assume that Γ_N is piecewise polyhedral and consists of

$$N = N_0 + \sum_{k=1}^{n_\Omega} N_k$$

plane triangular boundary elements. Here, N_0 denotes the number of boundary elements on the surface Γ_0 while N_k is the number of boundary elements on the surface Γ_k of the inclusion Ω_k, $k = 1, \ldots, n_\Omega$. The set of the nodes of the triangulation (5.25) will be denoted by $\{x_j\}_{j=1}^{M}$, where

$$M = M_0 + \sum_{k=1}^{n_\Omega} M_k \, .$$

We use the piecewise constant functions on Γ_N (ψ_ℓ is 1 on τ_ℓ and 0 outside τ_ℓ) as basis and test functions for the discretised single layer potentials. These functions also serve as test functions for the double layer potentials (5.13). The basis functions for the hypersingular and the adjoint double layer potentials are chosen to be piecewise linear: $\varphi_j(x_i) = \delta_{ij}$, φ_j is linear on each τ_ℓ. These functions are also used as trial functions for hypersingular operators (5.17) and double layer potentials (5.13).

Corresponding to the definition (5.15), the discretisation of the Steklov–Poincaré operators for a given subdomain is

$$S_h = D_h + \left(\frac{1}{2}M_h^\top + K_h^\top\right) V_h^{-1} \left(\frac{1}{2}M_h + K_h\right). \tag{5.26}$$

The above fully populated matrices are composed of the following three by three blocks

$$(V_h)_{k\ell} = \langle V(\psi_\ell I), \psi_k \rangle, \quad (K_h)_{kj} = \langle K(\varphi_j I), \psi_k \rangle, \quad (D_h)_{ij} = \langle D(\varphi_j I), \varphi_i \rangle,$$

where I is a three by three identity matrix, $\langle \cdot, \cdot \rangle$ denotes the scalar product on $L^2(\Gamma)$, and $k, \ell = 1, \ldots, N, i, j = 1, \ldots, M$. The sparse mass matrix M_h consists of blocks

$$(M_h)_{kj} = \left(\int_{\tau_k} \varphi_j(x) ds_x \right) I.$$

We renumber the degrees of freedom according to Cartesian directions. Thus, instead of composing the matrix V_h as N by N table of symmetric 3 by 3 blocks, we compose 6 symmetric N by N blocks. Furthermore, the matrix corresponding to the first term in (5.7) is computed separately. In these settings, the matrix V_h is

$$V_h = \frac{1}{2E} \left(\frac{1+\nu}{1-\nu} \right) \left((3 - 4\nu) \begin{pmatrix} V_{0,h} & 0 & 0 \\ 0 & V_{0,h} & 0 \\ 0 & 0 & V_{0,h} \end{pmatrix} + \begin{pmatrix} V_{11,h} & V_{21,h} & V_{13,h} \\ V_{21,h} & V_{22,h} & V_{23,h} \\ V_{31,h} & V_{32,h} & V_{33,h} \end{pmatrix} \right).$$

Here, the symmetric $N \times N$ matrix $V_{0,h}$ is the Galerkin discretisation of the single layer potential for the Laplace operator by the use of the piecewise constant functions. Its entries are given by

$$V_{0,h}[k, \ell] = \frac{1}{4\pi} \int_{\tau_k} \int_{\tau_\ell} \frac{1}{|x - y|} ds_y \, ds_x, \quad k, \ell = 1, \ldots, N. \quad (5.27)$$

The symmetric $N \times N$ matrices $V_{ij,h}$ are defined for $i, j = 1, 2, 3$ as follows

$$V_{ij,h}[k, \ell] = \frac{1}{4\pi} \int_{\tau_k} \int_{\tau_\ell} \frac{(x_i - y_i)(x_j - y_j)}{|x - y|^3} ds_y \, ds_x, \quad k, \ell = 1, \ldots, N. \quad (5.28)$$

These matrices do not depend on material parameters, and, therefore, can be repeatedly used for identical inclusions. The inner integrals in (5.27)–(5.28) can be computed analytically while the Gaussian quadrature rule should be applied to the outer integrals, see [45] for details.

The double layer potential operator K can be represented in terms of the single and double layer potential operators for the Laplace operator and of the above single layer potential operator for the Lamé system as follows

$$(Kv)(x) = (K_0 v)(x) - (V_0 M(\partial, n)v)(x) + \frac{E}{1+\nu}(V M(\partial, n)v)(x),$$

where K_0 and V_0 are the double and the single layer potentials for the Laplace operator, and V is the single layer potential of linear elasticity. In addition, we have used the matrix surface curl operator given by

$$M_{ij}(\partial_y, n(y)) = n_j(y)\frac{\partial}{\partial y_i} - n_i(y)\frac{\partial}{\partial y_j} \quad (5.29)$$

for $i, j = 1, 2, 3$. The Galerkin matrix $K_h \in \mathbb{R}^{3N \times 3M}$ for the double layer potential matrix of linear elasticity reads then

$$K_h = \begin{pmatrix} K_{0,h} & 0 & 0 \\ 0 & K_{0,h} & 0 \\ 0 & 0 & K_{0,h} \end{pmatrix} - \begin{pmatrix} V_{0,h} & 0 & 0 \\ 0 & V_{0,h} & 0 \\ 0 & 0 & V_{0,h} \end{pmatrix} \widetilde{T} + \frac{E}{1+\nu} V_h \widetilde{T},$$

where $V_{0,h}$ and $K_{0,h}$ are the Galerkin matrices related to the single and double layer potential of the Laplace operator. Furthermore, $\widetilde{T} \in \mathbb{R}^{3N \times 3M}$ is a sparse transformation matrix related to the matrix surface curl operator $M(\partial, \underline{n})$. The entries of the matrix $K_{0,h} \in \mathbb{R}^{N \times M}$ are

$$K_{0,h}[k, j] = \frac{1}{4\pi} \int\limits_{\tau_k} \int\limits_{\Gamma} \frac{(x-y, n(y))}{|x-y|^3} \varphi_j(y) ds_y ds_x$$

for $k = 1, \ldots, N$, $j = 1, \ldots, M$.

For the Galerkin discretisation of the hypersingular operator D, the following identity is used

$$\langle Du, v \rangle_\Gamma =$$

$$\frac{\mu}{4\pi} \iint\limits_{\Gamma \Gamma} \frac{1}{|x-y|} \sum_{k=1}^{3} \left(\frac{\partial}{\partial S_k(y)} u(y), \frac{\partial}{\partial S_k(x)} v(x) \right) ds_y ds_x +$$

$$\iint\limits_{\Gamma \Gamma} (M(\partial_x, n(x))v(x))^\top \left(\frac{\mu}{2\pi |x-y|} I - 4\mu^2 U^*(x, y) \right) M(\partial_y, n(y)) u(y) ds_y ds_x +$$

$$\frac{\mu}{4\pi} \iint\limits_{\Gamma \Gamma} \sum_{i,j,k=1}^{3} M_{kj}(\partial_x, n(x)) v_i(x) \frac{1}{|x-y|} M_{ki}(\partial_y, n(y)) v_j(y) ds_y ds_x$$

with the surface curl operator $M(\partial, n)$ defined by (5.29) and

$$\frac{\partial}{\partial S_1(x)} = M_{32}(\partial_x, n(x)), \quad \frac{\partial}{\partial S_2(x)} = M_{13}(\partial_x, n(x)), \quad \frac{\partial}{\partial S_3(x)} = M_{21}(\partial_x, n(x)).$$

Thus, the Galerkin matrix $D_h \in \mathbb{R}^{3M \times 3M}$ can be represented in terms of the transformation matrix $\widetilde{T}$ and of the Galerkin matrices related to the single layer potentials of both, the Laplace operator $V_{0,h}$ and the system of linear elastostatics V_h. We skip the details.

The fully populated matrices $V_{ij,h}$ and K_0, h needed for applying the discrete version of the Dirichlet-to-Neumann map S_h are efficiently approximated by the use of the Adaptive Cross Approximation procedure (ACA) (see [6, 8, 9]). This procedure reduces the computer memory requirement and the computational effort of the whole BEM method from quadratic to an almost linear complexity with respect to the number of nodes in the mesh.

5.3   Matrix Valued Radial Basis Functions

An interpolation by the means of radial basis functions (RBFs) has found its application in many different areas of research, e.g. data mining [29], imaging [2, 49],
neuronal networks [54] and machine learning [14]. The interpolant is sought as a
linear combination of radial functions, and the coefficients are found from the interpolation conditions. The procedure does not require an underlying mesh, and the
obtained interpolant is smooth. The solvability and the accuracy properties of the
interpolation problem have been extensively studied in [11, 15, 50].

Another application of the RBF technique is to approximate solutions of partial
differential equations (PDEs) [21] or BVPs [33, 50, 52, 53]. An approximate solution
$\tilde{u}$ to the scalar PDE of the form $Lu = f$ can be obtained by first constructing a RBF
interpolant for the right hand side f, i.e.

$$\sum_{j=1}^{N} \alpha_j \phi(x_i - x_j) = f(x_i), \quad X = \{x_j\}_{j=1}^{N} \subset \Omega \subset \mathbb{R}^d$$

$\alpha = (\alpha_1, \ldots, \alpha_N)^\top \in \mathbb{R}^N$ and then setting

$$\tilde{u}(x) = \sum_{j=1}^{N} \alpha_j \tilde{\phi}(x_i - x_j),$$

where the function $\tilde{\phi}$ satisfies $L\tilde{\phi} = \phi$. This approach was applied to the Laplacian
and Helmholtz-type operators in [12, 38, 48] and extended to other scalar operators
in [37, 46]. An application to the Lamé system for a special kind of the right hand
side can be found in [1]. More precisely, the volume force density of the right hand
side must originate form a scalar potential. Our aim is to extend the construction of
an approximate particular solution to cases with fairly general right hand sides. We
consider a system of PDEs and look for an approximation to its particular solution
as a linear combination of matrix-valued RBFs

$$u_p(x) = \sum_{j=1}^{N} \Phi(x - x_j)\alpha_j, \qquad (5.30)$$

where $\alpha_j \in \mathbb{R}^d$ and $\Phi : \mathbb{R}^d \to \mathbb{R}^d \times d$. First studies considering matrix-valued RBFs
to construct divergence free interpolants can be found in [35, 36, 40]. Later on, in
2008, a theory for curl-free interpolants was established in [19, 20].

We apply the operator L to the ansatz (5.30)

$$Lu_p(x) = \sum_{j=1}^{N} L\Phi(x - x_j)\alpha_j$$

and solve the interpolation problem

$$\sum_{j=1}^{N} L\Phi(x_i - x_j)\alpha_j = f(x_i), \quad i = 1, \ldots, N, \tag{5.31}$$

for a function $f : \Omega \to \mathbb{R}^d$. A major mathematical challenge in this approach is to specify the requirements on the operator L and the RBF Φ to guarantee the solvability of the system (5.31). In what follows, we will address these issues in the context of the Lamé system and refer to [23] for more details and proofs.

5.3.1 Functional Spaces

For $x \in \mathbb{R}^d$, we denote $\|x\|_2$ the Euclidean norm of x. Let the domain $\Omega \subset \mathbb{R}^d$ be bounded and simply connected having a $C^{k,1}$ boundary Γ. That is the boundary Γ is locally given by the graph of a $k - 1$ times continuously differentiable function and its derivatives of order k are Lipschitz continuous. $C^k(\Omega)$ is the space of k-times continuously differentiable functions. The Lebesgue spaces $L_p(\Omega)$ and the Sobolev spaces $W_p^k(\Omega)$ are defined in the usual way. On $\mathbb{R}^d$ for $p = 2$, we use the Fourier transform

$$\hat{f}(\xi) = \int_{\mathbb{R}^d} f(x) \exp(-ix^\top \xi) dx, \quad \xi \in \mathbb{R}^d$$

to characterise $H^\tau(\mathbb{R}^d)$ as

$$H^\tau(\mathbb{R}^d) = \left\{ f \in L_2(\mathbb{R}^d) : \hat{f}(\cdot)(1 + \|\cdot\|_2^2)^{\tau/2} \in L_2(\mathbb{R}^d) \right\}$$

The inverse Fourier transform is

$$f(x) = (2\pi)^{-d} \int_{\mathbb{R}^d} \hat{f}(\xi) \exp(i\xi^\top x) d\xi, \quad x \in \mathbb{R}^d.$$

Sobolev spaces for vector-valued functions $f : \Omega \to \mathbb{R}^d$ are equipped with norm

$$\|f\|_{(W_p^\tau(\Omega))^d} = \left(\sum_{j=1}^{d} \|f_j\|_{W_p^\tau(\Omega)}^p \right)^{1/p}, \quad 1 \le p < \infty,$$

and the semi-norm

$$|f|_{(W_p^\tau(\Omega))^d} = \left(\sum_{j=1}^{d} |f_j|_{W_p^\tau(\Omega)}^p \right)^{1/p}, \quad 1 \le p < \infty.$$

5.3.2 RBFs for the Lamé System

We recall the Lamé equations (5.5). In linear isotropic elastostatics the displacement field $u \in \mathbb{R}^3$ of an elastic body occupying some reference configuration $\Omega \subset \mathbb{R}^3$ satisfies the equilibrium equations or Lamé equations

$$Lu = -\mu \Delta u - (\lambda + \mu)\text{grad div} u = f \text{ in} \in \Omega .$$

The constants λ and μ are the so called Lamé constants and are described in form

$$\lambda = \frac{Ev}{(1+v)(1-2v)}, \quad \mu = \frac{E}{2(1+v)},$$

where $E > 0$ is the Young modulus and $v \in (0, 1/2)$ denotes the Poisson's ratio. For a given radial basis function $\phi : \mathbb{R}^3 \to \mathbb{R}$, we define the trivial matrix-valued RBF

$$\Phi_{\text{triv}} = \begin{pmatrix} \phi & 0 & 0 \\ 0 & \phi & 0 \\ 0 & 0 & \phi \end{pmatrix},$$

as an extension of the scalar valued case. Let $c \in \mathbb{R}^3$ be a constant vector. We can apply the differential operator L to the vector $\Phi_{\text{triv}}(x)c$ and obtain

$$L\Phi_{\text{triv}}(x)c = -\mu \Delta \Phi_{\text{triv}}(x)c - (\lambda + \mu)\text{grad div } \Phi_{\text{triv}}(x)c ,$$

which generates a new matrix-valued RBF $\Phi_{\text{Lamé}}$ having the components

$$(\Phi_{\text{Lamé}})_{ij} = -\delta_{ij}\mu \Delta \phi - (\lambda + \mu)\partial_{ij}\phi, \quad i, j = 1, 2, 3 .$$

An important role in the study of an RBF plays its positive definiteness.

Definition 5.1 A $d \times d$ matrix-valued function Φ is positive definite on $\mathbb{R}^d$ if for any set of distinct points $X = \{x_j\}_{j=1}^N \subset \mathbb{R}^d$ the quadratic form

$$\sum_{i,j=1}^N \alpha_i^\top \Phi(x_i - x_j)\alpha_j , \quad \alpha_j \in \mathbb{R}^d$$

is strictly positive for all $\alpha = (\alpha_1^\top, \ldots, \alpha_N^\top)^\top \neq 0$.

We will use the notation

$$A_{X,\Phi_{\text{Lamé}}} = \left(\Phi_{\text{Lamé}}(x_i - x_j)\right)_{i,j=1}^N \tag{5.32}$$

for the interpolation matrix.

Proposition 5.1 Let $\phi \in C^2(\mathbb{R}^3)$ a positive definite RBF in $\mathbb{R}^3$. Additionally assume ϕ and $-\Delta\phi$ are in $L_1(\mathbb{R}^3)$. Then the matrix-valued RBF $\Phi_{\text{Lamé}}$ is positive definite in the sense of Definition 5.1.

5.3.3 Native Spaces

Native spaces play a crucial role in error analysis and they will be used later for error estimates. Our presentation follows [19, Sect. 3.1], which is a simple generalisation of [50, Chap. 10].

Definition 5.2 Let H be a Hilbert space of vector-valued functions $f : \Omega \rightarrow \mathbb{R}^d$. A continuous $d \times d$ matrix-valued function Φ is called a reproducing kernel for H if for all $x \in \Omega$ and $c \in \mathbb{R}^d$ we have

1. $\Phi(\cdot - x)c \in H$,
2. $c^\top f(x) = (f, \Phi(\cdot - x)c)_H$ for all $f \in H$.

To construct the native space, we define an intermediate space

$$F_\Phi(\Omega) = \left\{ \sum_{j=1}^N \Phi(\cdot - x_j)\alpha_j : x_j \in \Omega, \alpha_j \in \mathbb{R}^d \text{ and } N \in \mathbb{N} \right\}$$

and equip this space with the bilinear form

$$\left(\sum_{j=1}^N \Phi(\cdot - x_j)\alpha_j, \sum_{i=1}^L \Phi(\cdot - y_i)\beta_i \right)_\Phi = \sum_{j=1}^N \sum_{i=1}^L \beta_i^\top \Phi(y_i - x_j)\alpha_j, \quad x_j, \ y_i \in \Omega.$$

This defines an inner product on $F_\Phi(\Omega)$ if Φ is symmetric positive definite. The completion of $F_\Phi(\Omega)$ with respect to the norm

$$\|\cdot\|_\Phi = \sqrt{(\cdot, \cdot)_\Phi}$$

is denoted by $\mathscr{F}_\Phi(\Omega)$. For an element $f \in \mathscr{F}_\Phi(\Omega)$, the values of the function f are defined by an injective linear map $R : \mathscr{F}_\Phi(\Omega) \rightarrow (C(\Omega))^d$ with

$$R(f)(x) = \left((f, \Phi(\cdot - x)e_1)_\Phi, \dots, (f, \Phi(\cdot - x)e_d)_\Phi \right)^\top.$$

Definition 5.3 The native space for a symmetric positive definite kernel Φ is defined by $\mathscr{N}_\Phi(\Omega) = R(\mathscr{F}_\Phi(\Omega))$ and is equipped with the inner product

$$(f, g)_{\mathscr{N}_\Phi(\Omega)} = (R^{-1}f, R^{-1}g)_{\mathscr{F}_\Phi(\Omega)}.$$

Lemma 5.1 *Suppose Φ is as in Definition 5.3. Further suppose $\mathscr{G}$ is a Hilbert space of functions $f : \Omega \to \mathbb{R}^d$ with reproducing kernel Φ. Then $\mathscr{G} = \mathscr{N}_\Phi(\Omega)$ and the inner products are the same.*

The theory for scalar-valued positive definite functions in [51] reveals an alternative representation.

Lemma 5.2 *Suppose that $\phi \in C(\mathbb{R}^d) \cap L_1(\mathbb{R}^d)$ is a real-valued positive definite function. Define*

$$
\mathscr{G} = \left\{ f \in L_2(\mathbb{R}^d) \cap C(\mathbb{R}^d) : \frac{\hat{f}}{\sqrt{\hat{\phi}}} \in L_2(\mathbb{R}^d) \right\}
$$

and equip this space with the bilinear form

$$
(f, g)_{\mathscr{G}} = (2\pi)^{-d} \left(\frac{\hat{f}}{\sqrt{\hat{\phi}}}, \frac{\hat{g}}{\sqrt{\hat{\phi}}} \right)_{L_2(\mathbb{R}^d)} = (2\pi)^{-d} \int_{\mathbb{R}^d} \frac{\overline{\hat{g}(\xi)}\,\hat{f}(\xi)}{\hat{\phi}(\xi)}\,\mathrm{d}\xi.
$$

Then $\mathscr{G}$ is a real Hilbert space with inner product $(\cdot, \cdot)_{\mathscr{G}}$ and reproducing kernel ϕ. Hence $\mathscr{G}$ is the native space of ϕ on $\mathbb{R}^d$, i.e. $\mathscr{G} = \mathscr{N}_\phi(\mathbb{R}^d)$, and both inner products coincide. In particular, every $f \in \mathscr{N}_\phi(\mathbb{R}^d)$ can be recovered from its Fourier transform $\hat{f} \in L_1(\mathbb{R}^d) \cap L_2(\mathbb{R}^d)$.

In [19], a generalisation of Lemma 5.2 by the pseudo-inverse is given. This construction is necessary, since the Fourier transforms of Φ_{div} and Φ_{curl} are not invertible. But in our case, the Fourier transform is always invertible. So we can generalise Lemma 5.2 as

Lemma 5.3 *Suppose that $\phi \in C^2(\mathbb{R}^3) \cap L_1(\mathbb{R}^3)$ is a real-valued positive definite function and $-\Delta\phi \in L_1(\mathbb{R}^3)$. Define*

$$
\mathscr{G}_{\mathrm{Lam\acute{e}}} = \left\{ f \in (C(\mathbb{R}^3) \cap L_1(\mathbb{R}^3))^3 : \int_{\mathbb{R}^3} \hat{f}(\xi)^* \widehat{\Phi_{\mathrm{Lam\acute{e}}}}(\xi)^{-1} \hat{f}(\xi)\,\mathrm{d}\xi < \infty \right\}
$$

and equip this space with the bilinear form

$$
(f, g)_{\mathscr{G}_{\mathrm{Lam\acute{e}}}} = (2\pi)^{-3} \int_{\mathbb{R}^3} \hat{g}(\xi)^* \widehat{\Phi_{\mathrm{Lam\acute{e}}}}(\xi)^{-1} \hat{f}(\xi)\,\mathrm{d}\xi
$$

Then $\mathscr{G}_{\mathrm{Lam\acute{e}}}$ is a real Hilbert space with inner product $(\cdot, \cdot)_{\mathscr{G}_{\mathrm{Lam\acute{e}}}}$ and reproducing kernel $\Phi_{\mathrm{Lam\acute{e}}}$. Hence $\mathscr{G}$ is the native space of $\Phi_{\mathrm{Lam\acute{e}}}$ on $\mathbb{R}^3$, i.e. $\mathscr{G} = \mathscr{N}_{\Phi_{\mathrm{Lam\acute{e}}}}(\mathbb{R}^3)$, and both inner products coincide.

5.3.4 Stability

Following the works of Wendland and Fuselier, [19, 50], we look for a function Ψ for which

$$\sum_{j,k=1}^{N} \alpha_j^\top \Phi(x_j - x_k)\alpha_k \geq \sum_{j,k=1}^{N} \alpha_j^\top \Psi(x_j - x_k)\alpha_k \geq \theta \, \|\alpha\|_2^2 \text{ for all } \alpha_j \in \mathbb{R}^3$$

holds. Then θ is a lower bound for the smallest eigenvalue of the interpolation matrix $A_{X,\Phi_{\text{Lamé}}}$. To get a simple condition on Ψ, we first take a look at the quadratic form

$$\alpha^\top A_{X,\Phi_{\text{Lamé}}} \alpha =$$

$$\int_{\mathbb{R}^3} \left| \sqrt{\lambda + \mu} \sum_{i=1}^{N} \xi^\top \alpha_i \exp(-ix_i^\top \xi) \right|^2 \hat{\phi}(\xi) d\xi +$$

$$\int_{\mathbb{R}^3} \|\xi\|_2^2 \left\| \sqrt{\mu} \sum_{i=1}^{N} \alpha_i \exp(-ix_i^\top \xi) \right\|_2^2 \hat{\phi}(\xi) d\xi .$$

For a positive definite function ϕ, we know according to [50], that $\hat{\phi}$ is positive. It suffices now to find a positive $\hat{\psi}$ with $\hat{\phi} \geq \hat{\psi}$. In [39], a general method for constructing ψ is proposed, but it suffices here to choose a simpler one. Define the characteristic function of the ball with radius $\sigma/2$ in d dimensions, centred at $0 \in \mathbb{R}^d$ with $\chi_{\frac{\sigma}{2},d}$. According to [50, Chap. 12], the inverse Fourier transform is given by

$$\check{\chi}_{\frac{\sigma}{2},d}(x) = \left(\frac{\sigma}{4\pi \, \|x\|_2} \right)^{d/2} J_{\frac{d}{2}} \left(\frac{\|x\|_2 \, \sigma}{2} \right)$$

where $J_{\frac{d}{2}}$ is a Bessel function of the first kind.

Lemma 5.4 *Define* $\psi_\sigma = \check{\chi}_{\frac{\sigma}{2},d}^2$ *and the d-dimensional matrix valued RBF*

$$(\Psi_{\text{Lamé}})_{ij} = -\delta_{ij} \mu \Delta \psi_\sigma - (\lambda + \mu) \partial_{ij} \psi_\sigma , \ i, j = 1, \ldots, d .$$

Then $\Psi_{\text{Lamé}}$ *has the form*

$$\Psi_{\text{Lamé}}(x) = -B(x)((\lambda + \mu)xx^\top + d\mu \, \|x\|_2^2 I) - A(x)(\lambda + (d+1)\mu)I,$$

where

$$B(x) = 8\pi^2 \left(\left(\check{\chi}_{\frac{\sigma}{2},d+2}(x) \right)^2 + \check{\chi}_{\frac{\sigma}{2},d}(x) \cdot \check{\chi}_{\frac{\sigma}{2},d+4}(x) \right),$$
$$A(x) = -4\pi \, \check{\chi}_{\frac{\sigma}{2},d}(x) \cdot \check{\chi}_{\frac{\sigma}{2},d+2}(x) .$$

The eigenvalues of $\Psi_{\text{Lamé}}$ are given by

$$
\begin{cases}
-\mu d \, \|x\|_2^2 \, B(x) - A(x)(\lambda + (d+1)\mu), & \text{multiplicity } d-1, \\
-(B(x) \, \|x\|_2^2 + A(x))(\lambda + (d+1)\mu), & \text{multiplicity } 1.
\end{cases}
$$

Lemma 5.5 *It holds*

$$
|A(x)| \leq 2^{d+5} \left(\frac{\sigma^2}{8\pi} \right)^{d+1} \left(\frac{\|x\|_2 \, \sigma}{2} \right)^{-d-2},
$$

$$
\|x\|_2^2 \, |B(x)| \leq 2^{d+7} \left(\frac{\sigma^2}{8\pi} \right)^{d+1} \left(\frac{\|x\|_2 \, \sigma}{2} \right)^{-d-1}.
$$

We now establish an upper bound for the largest eigenvalue of $\Psi_{\text{Lamé}}$

$$
\Lambda(x) = -\left(B(x) \, \|x\|_2^2 + A(x) \right)\left(\lambda + (d+1)\mu \right).
$$

Corollary 5.1 *It holds*

$$
\Lambda(x) \leq (\lambda + (d+1)\mu) 2^{d+5} \left(\frac{\sigma^2}{8\pi} \right)^{d+1} \left(\left(\frac{\|x\|_2 \, \sigma}{2} \right)^{-d-2} + 4 \left(\frac{\|x\|_2 \, \sigma}{2} \right)^{-d-1} \right).
$$

The Lemma from [40, 50, Chap. 11] reads:

Lemma 5.6 *Let $f : \mathbb{R} \to \mathbb{R}$ be a scalar valued function and $X \subset \mathbb{R}^d$ a set of points and $q_X = \frac{1}{2} \min_{i \neq j} \|x_i - x_j\|_2$ its separation distance. Then*

$$
\sum_{j \neq k} f(\|x_j - x_k\|_2) \leq 3^d \sum_{m=1}^{\infty} m^{d-1} \kappa_{f,m},
$$

where

$$
\kappa_{f,m} = \sup \{ |f(\|x\|_2)| : \ m q_X \leq \|x\|_2 \leq (m+1) q_X \}.
$$

Lemma 5.7 *Let*

$$
\sigma \geq \max \left\{ \frac{2}{q_X}, \frac{\tilde{C}}{q_X} \right\}, \quad \tilde{C} = \left(\frac{2d(d+2)5\pi}{8(d-1)} \Gamma^2 \left(\frac{d+2}{2} \right) \right)^{1/(d+1)}.
$$

We have

$$
\max_k \sum_{j \neq k} \Lambda(x_j - x_k) \leq \frac{-(\lambda + (d+1)\mu)}{2d} A(0).
$$

Theorem 5.1 *Let ϕ be an even conditionally positive definite function that possesses a positive Fourier transform $\hat{\phi} \in C(\mathbb{R}^d)$. With the function*

$$M(\sigma) = \inf_{\|\xi\|_2 \leq \sigma} \hat{\phi}(\xi).$$

and under the premises of Lemmata 5.1 and 5.7 a lower bound on $\lambda_{\min}(A_{X,\Phi_{\text{Lamé}}})$ is given by

$$\lambda_{\min}(A_{X,\Phi_{\text{Lamé}}}) \geq \frac{2}{3}\mu \left(\frac{\sigma^2}{16\pi}\right)^{(d+2)/2} \frac{M(\sigma)4\pi}{(4\pi)^d \, \Gamma(d/2+1)},$$

where $\sigma \geq \tilde{C}/q_X$ and a constant $\tilde{C}$ is independent of λ, μ, ϕ and X.

5.3.5   Error Estimates

To identify the native space with a Sobolev space, it does not suffice that ϕ is positive definite. Additionally, we assume that an algebraic decay condition

$$c_1(1 + \|\xi\|_2^2)^{-(s+1)} \leq \hat{\phi}(\xi) \leq c_2(1 + \|\xi\|_2^2)^{-(s+1)} \tag{5.33}$$

holds for $\hat{\phi}$ and some $s \in \mathbb{R}$. The domain Ω fulfils the properties that are required to use the extension operator $\mathfrak{E}$ from Stein, [47]. First we define an intermediate space $\tilde{H}^s(\mathbb{R}^3)$ by

$$(\tilde{H}^s(\mathbb{R}^3))^3 = \left\{ f \in (L_2(\mathbb{R}^3)^3 : \int_{\mathbb{R}^3} \frac{\left\|\hat{f}(\xi)\right\|_2^2}{\|\xi\|_2^2} (1 + \|\xi\|_2^2)^{s+1} \, d\xi < \infty \right\}. \tag{5.34}$$

Proposition 5.2 *If the algebraic decay condition holds for ϕ, then the norms on $(\tilde{H}^s(\mathbb{R}^3))^3$ and $\mathcal{G}_{\text{Lamé}}$ are equivalent.*

To identify $(\tilde{H}^s(\mathbb{R}^3))^3$ as subspace of $(H^s(\mathbb{R}^3))^3$, we need an additional Lemma.

Lemma 5.8 ([18]). *In the space $(\tilde{H}^s(\mathbb{R}^3))^3$, the norm (5.34) is equivalent to the norm defined by*

$$\|f\|_*^2 = \int_{\mathbb{R}^3} \frac{\left\|\hat{f}(\xi)\right\|_2^2}{\|\xi\|_2^2} \, d\xi + \|f\|_{(H^s(\mathbb{R}^3))^3}^2.$$

Thus, the space $(\tilde{H}^s(\mathbb{R}^3))^3$ is a subspace of $(H^s(\mathbb{R}^3))^3$. To get error estimates for functions in $H^s(\Omega)$, we need an extension from $H^s(\Omega)$ to $\tilde{H}^s(\mathbb{R}^3)$. For this purpose we use Stein's extension operator $\mathfrak{E}$, [47], and therefore Ω has to be bounded and its surface have to be smooth enough. We define the extension operator $\tilde{\mathfrak{F}}$ by

$$\tilde{\mathfrak{F}}g(x) = \mathfrak{E}g(x) - \mathfrak{E}g(x - x_0). \tag{5.35}$$

The point x_0 must be chosen on such a way that the supports of $\mathfrak{E}g$ and $\mathfrak{E}g(\cdot - x_0)$ do not intersect. This is possible due to the construction of Stein's extension operator. Then the operator $\tilde{\mathfrak{F}}$ defines an extension

Let $g \in (H^s(\Omega))^3$. We have $g \in (L_1(\mathbb{R}^3))^3$, because $g \in (L_2(\mathbb{R}^3))^3$ has a compact support, as does $\tilde{\mathfrak{F}}g$. Consequently $\widehat{\tilde{\mathfrak{F}}g}$ is continuous and $\widehat{\tilde{\mathfrak{F}}g}(0)$ is well defined. Due to the construction it follows $\widehat{\tilde{\mathfrak{F}}g}(0) = 0$. Then with

$$\left| \widehat{\tilde{\mathfrak{F}}g}(\xi) - 0 \right| = \left| \widehat{\tilde{\mathfrak{F}}g}(\xi) - \widehat{\tilde{\mathfrak{F}}g}(0) \right| = \left| \int\limits_{\substack{\text{supp}(g) \subset \mathbb{R}^3 \\ \text{compact}}} \tilde{\mathfrak{F}}g(x)(1 - \exp(-\iota \xi^\top x)) dx \right|$$

$$\leq \int\limits_{\substack{\text{supp}(g) \subset \mathbb{R}^3 \\ \text{compact}}} |\tilde{\mathfrak{F}}g(x)| \, \|\xi\|_2 \, \|x\|_2 \, dx \leq \|\xi\|_2 \left\| \tilde{\mathfrak{F}}g(x) \right\|_{L_1} \sup_{\substack{x \in \text{supp}(g) \subset \mathbb{R}^3 \\ \text{compact}}} \|x\|_2$$

for $\|\xi\|_2 \leq 1$, we have

$$\left\| \tilde{\mathfrak{F}}g \right\|_{(\tilde{H}^s(\Omega))^3} = \int\limits_{\mathbb{R}^3} \frac{\left\| \widehat{\tilde{\mathfrak{F}}g}(\xi) \right\|_2^2}{\|\xi\|_2^2} d\xi + \left\| \tilde{\mathfrak{F}}g \right\|_{(H^s(\mathbb{R}^3))^3}^2$$

$$= \int\limits_{\|\xi\| \leq 1} \frac{\left\| \widehat{\tilde{\mathfrak{F}}g}(\xi) \right\|_2^2}{\|\xi\|_2^2} d\xi + \int\limits_{\|\xi\| > 1} \frac{\left\| \widehat{\tilde{\mathfrak{F}}g}(\xi) \right\|_2^2}{\|\xi\|_2^2} d\xi + \left\| \tilde{\mathfrak{F}}g \right\|_{(H^s(\mathbb{R}^3))^3}^2$$

$$\leq \int\limits_{\|\xi\| \leq 1} \frac{\left\| \widehat{\tilde{\mathfrak{F}}g}(\xi) \right\|_{L_1}^2 \|\xi\|_2^2}{\|\xi\|_2^2} d\xi + \left\| \tilde{\mathfrak{F}}g \right\|_{(H^s(\mathbb{R}^3))^3}^2 + \left\| \tilde{\mathfrak{F}}g \right\|_{(H^s(\mathbb{R}^3))^3}^2$$

$$\leq C \left\| \tilde{\mathfrak{F}}g \right\|_{(H^s(\mathbb{R}^3))^3}^2 .$$

As a direct consequence and Lemma 5.8, we can identify functions $H^s(\Omega)$ as the restriction of functions $\tilde{H}^s(\mathbb{R}^3)$ to Ω.

Proposition 5.3 *Let $\Omega \subset \mathbb{R}^3$ be a bounded domain with a smooth boundary. Then it follows*

$$H^s(\Omega) = \left\{ f|_\Omega : \ f \in \tilde{H}^s(\mathbb{R}^3) \right\}$$

and the norms are equivalent.

To get interpolation error estimates for RBF outside the native space, we first mention the result of Wendland, Narcowich and Ward [41, 51] concerning the error estimates in Sobolev spaces for functions with many zeros. This result will be applied later to the error $f - s_{f,X}$.

Proposition 5.4 *Suppose Ω is bounded and satisfies an interior cone condition. Let k be a positive integer, $0 < s \le 1$, $1 \le p < \infty$, and let $m \in \mathbb{N}_0$ satisfy $k > m + d/p$ for $p > 1$ or, for $p = 1$, $k \ge m + d$. Also let $X \subset \Omega$ be a discrete set with sufficiently small mesh norm h_X sufficient small. If $u \in W_p^{k+s}(\Omega)$ satisfies $u|_X = 0$ then*

$$|u|_{W_q^{|\alpha|}(\Omega)} \le C_{k,d,p,q,|\alpha|} h_{X,\Omega}^{k+s-|\alpha|-d(1/p-1/q)_+} |u|_{W_p^{k+s}(\Omega)},$$

where $(x)_+ = x$ if $x \ge 0$ and is 0 otherwise and the semi norm

$$|u|_{W_q^k(\Omega)} = \left(\sum_{|\alpha|=k} \|D^\alpha u\|_{L_q(\Omega)}^q \right)^{1/q}.$$

Remark 5.1 In the vector valued case, we have the same result for functions $u \in (W_p^{k+s}(\Omega))^3$.

We can now apply the method from [20] and [41] to get interpolation error estimates for functions outside the native space. To do so, we first mention a few facts about band limited interpolation and the reproducing kernel $\mathcal{K}^\tau$.

Band-Limited Functions

In this subsection we review and establish important approximation properties of band-limited function. Let $\sigma > 0$ and define $\mathscr{B}_\sigma$ to be

$$\mathscr{B}_\sigma = \left\{ f \in (L_2(\mathbb{R}^3))^3 : \operatorname{supp}(\hat{f}) \in B(0,\sigma) \right\}$$

where $B(0,\sigma)$ is the ball in $\mathbb{R}^3$ having centre in 0 and of the radius σ. We need a further space

$$\tilde{\mathscr{B}}_\sigma = \left\{ f \in \mathscr{B}_\sigma : \int_{\mathbb{R}^3} \frac{\left\| \hat{f}(\xi) \right\|_2^2}{\|\xi\|_2^2} d\xi < \infty \right\},$$

which can be seen according to Lemma 5.8 as the subspace of band-limited functions in $(\tilde{H}^s(\mathbb{R}^3))^3$. Let $f \in \tilde{H}^r(\mathbb{R}^3)$. Then we can define a band-limited function $\hat{g}_\sigma = \hat{f}\chi_\sigma$, where χ_σ is the characteristic function of the ball $B(0,\sigma)$ and formulate the following lemma:

Lemma 5.9 *Let $t \ge r \ge 0$. For every function $f \in \tilde{H}^r(\mathbb{R}^3)$ and every $\sigma > 0$ exists a function $g_\sigma \in \tilde{\mathscr{B}}_\sigma$ with the approximation property*

$$\|f - g_\sigma\|_{\tilde{H}^r} \le \sigma^{r-t} \|f\|_{\tilde{H}^t}.$$

Additionally, we need some properties of the space $\tilde{H}^\tau(\mathbb{R}^3)$. This space is a Hilbert space with reproducing kernel $\mathscr{K}^\tau$ for $\tau > 3/2$. We define the kernel $\mathscr{K}^\tau$ via its Fourier transform and to identify the space later properly, we assume that (5.33) holds. Additionally, the following results are independent of the chosen ϕ.

We define

$$\hat{\mathscr{K}}^\tau(\xi) = (\mu \,\|\xi\|_2^2 \, I + (\lambda + \mu)\xi\xi^\top)(1 + \|\xi\|_2^2)^{-(\tau+1)}.$$

The inverse Fourier transform of this function is given by

$$\mathscr{K}^\tau(x) = c_\tau(-\mu\Delta I - (\lambda + \mu)\nabla\nabla^\top)\,\|x\|_2^{\tau-1/2}\,K_{\tau-1/2}(\|x\|_2),$$

where K is the modified Bessel function of the second kind and c_τ is a known constant (see [50, Theorem 6.13]). We apply the theory from Sect. 5.3.4 to the kernel $\mathscr{K}^\tau$ and consider interpolants of the form

$$g(x) = \sum_{j=1}^N \mathscr{K}^\tau(x - x_j)c_j, \quad c_j \in \mathbb{R}^3, \,, j = 1, \ldots, N$$

on the set $X = \{x_1, \ldots, x_N\}$. Thus

$$\|g\|_{(\tilde{H}^\tau(\mathbb{R}^3))^3}^2 = (g, g)_{\mathscr{G}_{\text{Lamé}}} = \left(\sum_{i=1}^N \mathscr{K}^\tau(x - x_i)c_i, \sum_{j=1}^N \mathscr{K}^\tau(x - x_j)c_j\right)_{\mathscr{G}_{\text{Lamé}}}$$

$$= (2\pi)^{-3}\int_{\mathbb{R}^3} \left(\sum_{j=1}^N \hat{\mathscr{K}}^\tau(\xi)c_j\, e^{\imath\, x_j^\top \xi}\right)^* \left(\sum_{i=1}^N \hat{\mathscr{K}}^\tau(\xi)^{-1}\hat{\mathscr{K}}^\tau(\xi)c_i\, e^{\imath\, x_i^\top \xi}\right)\,d\xi$$

$$= (2\pi)^{-3}\sum_{i,j=1}^N \int_{\mathbb{R}^3} \left(\hat{\mathscr{K}}^\tau(\xi)c_j\right)^* c_i\, e^{-\imath\,(x_j-x_i)^\top \xi}\,d\xi = \sum_{i,j=1}^N c_j^\top \mathscr{K}^\tau(x_j - x_i)c_i.$$

This is exactly the bilinear form from the stability estimates. It follows

$$\lambda_{\min}(A_{X,\mathscr{K}^\tau})\,\|c\|_2^2 \leq \|g\|_{(\tilde{H}^\tau(\mathbb{R}^3))^3}^2 \leq \lambda_{\max}(A_{X,\mathscr{K}^\tau})\,\|c\|_2^2,$$

where $c = (c_1^\top, \ldots, c_N^\top)^\top \in \mathbb{R}^{3N}$.

Lemma 5.10 *The smallest eigenvalue can be estimated by*

$$\lambda_{\min}(A_{X,\mathscr{K}^\tau}) \geq c_{\tau,\lambda,\mu}q_X^{2\tau-3}.$$

Lemma 5.11 *The kernel* $\mathcal{K}^\tau$ *has the following explicit form*

$$\mathcal{K}^\tau(x) = c_\tau \left(-\mu \left(\|x\|_2^{\nu-1} \left(-3K_{\nu-1}(\|x\|_2) + K_{\nu-2}(\|x\|_2) \|x\|_2 \right) \right) I \right.$$
$$\left. - (\lambda + \mu) \|x\|_2^{\nu-2} K_{\nu-2}(\|x\|_2) xx^\top \right),$$

where $\nu = \tau - 1/2$. *The eigenvalues of the kernel* $\mathcal{K}^\tau$ *are*

$$-c_\tau \mu \left(\|x\|_2^{\nu-1} \left(-3K_{\nu-1}(\|x\|_2) + K_{\nu-2}(\|x\|_2) \|x\|_2 \right) \right)$$

which is double and the single eigenvalue

$$-c_\tau \mu \left(\|x\|_2^{\nu-1} \left(-3K_{\nu-1}(\|x\|_2) + K_{\nu-2}(\|x\|_2) \|x\|_2 \right) \right) - c_\tau (\lambda + \mu) \|x\|_2^\nu K_{\nu-2}(\|x\|_2).$$

The single eigenvalue has the largest absolute value, which decreases for $\|x\|_2 > 3$. *Furthermore it holds* $0 < \mathcal{K}^\tau(0) < \infty$.

To get interpolation error estimates for band-limited functions, we follow [42].

Lemma 5.12 *Let*

$$g = \sum_{j=1}^{N} \mathcal{K}^\tau(\cdot - x_j) c_j, \quad c_j \in \mathbb{R}^3$$

and define g_σ *by* $\hat{g}_\sigma = \hat{g}\chi_\sigma$. *Then there exists a constant* $\kappa > 0$, *which is independent of* $X = (x_1, \ldots, x_N)$ *and the* c_j's, *such that for* $\sigma = \kappa/q_X$ *the inequality*

$$I_\sigma = \|g - g_\sigma\|_{(\tilde{H}^\tau(\mathbb{R}^3))^3} \le \frac{1}{2} \|g\|_{(\tilde{H}^\tau(\mathbb{R}^3))^3}$$

holds.

The following theorem gives a bound on the distance between a given function $f \in (\tilde{H}^\tau(\mathbb{R}^3))^3$ and $f_\sigma \in \tilde{\mathcal{B}}_\sigma$. The estimate is exactly what one can expect according to [20, Theorem 1] and [42, Theorem 3.4].

Proposition 5.5 *Let* $\tau, t \in \mathbb{R}$ *such that* $\tau > 3/2$ *and* $t \ge 0$. *Let* $X = \{x_j\}_{j=1}^N \subset \mathbb{R}^3$ *be a point set with separation distance* q_X. *If* $f \in (\tilde{H}^{\tau+t}(\mathbb{R}^3))^3$, *then there exists a function* $f_\sigma \in \tilde{\mathcal{B}}_\sigma(\mathbb{R}^3)$ *such that* $f_\sigma|_X = f|_X$ *and*

$$\|f - f_\sigma\|_{(\tilde{H}^\tau(\mathbb{R}^3))^3} \le 5 \, \mathrm{dist}_{(\tilde{H}^\tau(\mathbb{R}^3))^3}(f, \tilde{\mathcal{B}}_\sigma) \le 5\kappa^{-t} q_X^t \|f\|_{(\tilde{H}^{\tau+t}(\mathbb{R}^3))^3}$$

with $\sigma = \kappa/q_X$, *where* $\kappa \ge 1$ *depends only on* τ, λ *and* μ.

The following result estimates the RBF approximation error for functions outside the native space.

Proposition 5.6 *Let k and s be as in Proposition 5.4 and τ decomposed as $\tau = k + s$. Further let $3/2 < \beta \leq \tau$ be a positive integer. If $f \in (H^{\beta}(\Omega))^3$ then*

$$\left\| f - s_{f,X} \right\|_{(H^{\mu}(\Omega))^3} \leq C \, h_{X,\Omega}^{\beta-\mu} \rho_{X,\Omega}^{\tau-\beta} \left\| f \right\|_{(H^{\beta}(\Omega))^3}$$

for all $0 \leq \mu \leq \beta$.

5.4 Numerical Methods for Matrix Valued RBFs

With the existence, stability, and accuracy results at hand, we turn to the task of computing the interpolation coefficients and evaluating the interpolant. More precisely, we formulate an efficient numerical algorithm to solve the linear system (5.31) with the matrix valued RBF given by (5.32).

It is easy to see, that our RBFs do not have local supports, thus the resulting system matrix is fully populated. This implies, that a direct solution of the interpolation problem with N unknowns requires $\mathcal{O}(N^3)$ floating point operations. To alleviate these prohibitively high costs we propose an iterative solution scheme based on the Krylov subspace method (see Sect. 5.4.2). It requires repeated computations of system matrix-vector products, which cost $\mathcal{O}(N^2)$ operations. A result of any such matrix-vector product, however, can be approximately computed with a desired accuracy with an almost linear cost. In our study, this acceleration is a byproduct of a block-wise low rank approximation of the system matrix (see Sect. 5.4.1). This approximation is constructed by means of the H- or H^2-matrix technique. The same technique is also employed during the evaluation of the interpolant to reduce the complexity.

5.4.1 Fast Matrix Vector Multiplication and Evaluation

A number of ways to accelerate matrix-vector products in the context of RBF interpolation have been proposed in recent years. In [43], certain optimality properties of the thin plate splines were used to develop a method for fast matrix vector multiplication in the context of the RBF interpolation. In [2], the symmetries of a tensor grid were utilised to rapidly evaluate the RBF interpolant. This particular choice of the grid also allows for a special evaluation procedure via FFT of recursive Toeplitz matrices (see [31]).

Another approach can be found in [4]. The authors recognised similarities between the N-body problem and the evaluation of thin-plate splines and formulated a Fast-Multipole method for the RBF interpolation. This led to evaluation procedures with complexity of order $\mathcal{O}(N \log N)$ for various RBFs (see [3, 13, 44]).

One of the recently proposed methods is the multilevel evaluation, [34], a procedure with the complexity of order $\mathcal{O}(N)$. In numerical tests, the order $\mathcal{O}(N)$ is observed for the two- and three-dimensional case. However, the extension of this method to the matrix-valued case is not possible, since it relies on the radial symmetry of the functions.

In this study we employ a rapid interpolation and evaluation method based on hierarchical, block-wise low rank matrices. Applications of this technique to scalar RBFs can be found in [7, 22].

The H-matrix approximation technique was first presented in [24] as a generalisation of the panel clustering method. In general, if a given matrix can be partitioned into low rank blocks it is referred to as an H-matrix. In practical applications the structure of an H-matrix is based block clusters trees. The latter comes from a hierarchical decomposition of the underlying discrete geometry (e.g. point sets, triangular mesh).

The detailed description of construction of H-matrix approximants to the fully populated matrices arising in numerical methods for elliptic boundary value problems can be found in [27]. These approximants can be constructed with a chosen accuracy. The numerical cost and the memory requirement for the construction are almost linear with respect to the number of rows/columns. Recipes for efficient operations with H-matrices, such as multiplication, approximate inversion, factorisation can be found in [7] and in [26]. In our study, we use these results to accelerate operations with the BEM matrices (5.27), (5.30).

A further development of the H-matrix technique, the H^2-matrices, are presented in [28]. The introduction of the so-called cluster bases allows for reducing the complexity order to the linear one. An efficient way to perform linear algebra operations with H^2-matrices was proposed in [7, 10, 25, 28].

An H- or an H^2-matrix approximant to the system matrix in our interpolation problem (5.31) can be constructed if the element generating function fulfils certain conditions. The matrix $(\Phi_{\text{Lamé}})_{ij}$, $i, j = 1, 2, 3$ consists of the partial derivatives of the kernel function ϕ. The key requirement is that this function must be asymptotically smooth.

Definition 5.4 ([7]). A function $\kappa : \Omega \times \mathbb{R}^n \to \mathbb{R}$ satisfying $\kappa(x, \cdot) \in C^\infty(\mathbb{R}^n \backslash \{x\})$ for all $x \in \Omega$ is called asymptotically smooth in Ω with respect to y if constants c and γ can be found such that for all $x \in \Omega$ and all $\alpha \in \mathbb{N}_0^n$

$$|\partial_y^\alpha \kappa(x, y)| \leq c|\alpha|!\gamma^{|\alpha|} \frac{|\kappa(x, y)|}{\|x - y\|_2^{|\alpha|}} \quad \text{for all } y \in \mathbb{R}^n \backslash \{x\}.$$

This requirement is fulfilled, for example, by the Wendland function or inverse multiquadrics (see [30] for more details).

5.4.2 Fast Solver

The number of fast solvers for the interpolation problem is quite limited in comparison to the algorithms for fast matrix-vector multiplication. They are, in particular, an iterative method from [5] and the Countour-Padé or RBF-QR algorithm, [17].

Here, we extend a Krylov method developed by Faul and Powell, [16], to the case of matrix-valued RBFs. The method requires the underlying operator to be self-adjoint, nevertheless it can be adapted for handling positive-semi-definite RBFs.

Induced Norm

Let s be of the form

$$s(x) = \sum_{j=1}^{N} \Phi(x - x_j)c_j, \tag{5.36}$$

with coefficients $c_j \in \mathbb{R}^3$ and $X = \{x_1, \ldots, x_N\} \subset \mathbb{R}^3$. Define the matrix M by

$$M = (\Phi(x_i - x_j))_{i,j=1,\ldots N} \in \mathbb{R}^{3N \times 3N}$$

which consist of blocks of the form

$$(\Phi_{kl}(x_i - x_j))_{k,l=1,2,3}$$

for $i, j = 1, \ldots, N$. The coefficient vector c is defined by

$$c = (c_1^\top, \ldots, c_N^\top)^\top \in \mathbb{R}^{3N}.$$

The interpolation problem is then given by

$$Mc = f,$$

for given point evaluations $f(x_i) \in \mathbb{R}^3$, $i = 1, \ldots N$. We denote the space of all functions s of the form (5.36) by $\mathcal{S}$. For two functions $s, t \in \mathcal{S}$,

$$t(x) = \sum_{i=1}^{N} \Phi(x - x_j)d_j,$$

we define a bilinear form on $\mathcal{S}$ by

$$\langle s, t \rangle = c^\top M d.$$

By the virtue of the positive definiteness of Φ we obtain a norm

$$\|s\|_{\mathcal{S}} = \langle s, s \rangle^{1/2}.$$

Corollary 5.2 *For $s, t \in \mathcal{S}$, it follows*

$$\langle s, t \rangle = \sum_{i=1}^{N} c_i^\top t(x_i) = \sum_{i=1}^{N} s(x_i)^\top d_j.$$

Proof A simple calculation shows

$$\langle s, t \rangle = \sum_{i=1}^{N} c_j^\top \sum_{j=1}^{N} \Phi(x_i - x_j) d_j = \sum_{i=1}^{N} c_j^\top t(x_i) = \sum_{i=1}^{N} s(x_i)^\top d_j.$$

□

Thus, the matrix M is not needed to evaluate the bilinear form. Define the required interpolant by $s^\star$. One can evaluate

$$\langle s, s^\star \rangle = \sum_{i=1}^{N} c_j^\top s^\star(x_i) = \sum_{i=1}^{N} c_j^\top f(x_i)$$

without knowing $s^\star$.

Krylov Subspace Method

The Krylov subspace method is an iterative solution method for the linear discrete problem. It is based on the properties of the underlying linear operator $A : \mathscr{S} \to \mathscr{S}$. Let k be the iteration number, and $\mathscr{S}_k$ is the linear subspace of $\mathscr{S}$, which is spanned by $A^j s^\star$, $j = 1, ..., k$, where $s^\star$ is the required interpolant. The aim is to find an interpolant s_{k+1} with

$$\left\| s^\star - s_{k+1} \right\|_{\mathscr{S}} < \left\| s^\star - s_k \right\|_{\mathscr{S}}. \tag{5.37}$$

The sequence $\left\| s^\star - s_j \right\|_{\mathscr{S}}$, $j = 1, ...,$ decreases strictly monotonically. The coefficients $A^j s^\star$ however are not computed directly due to numerical instability. Instead, we define search directions d_k to find s_{k+1} with a scaling factor α_k by

$$s_{k+1} = s_k + \alpha_k d_k, \quad \alpha_k \in \mathbb{R}. \tag{5.38}$$

The search directions have to fulfil the orthogonality condition $\langle d_k, d_{k-1} \rangle = 0$. We can compute d_k by

$$d_k = A(s^\star - s_k) + \beta_k d_{k-1}, \quad \beta_k \in \mathbb{R}. \tag{5.39}$$

The minimisation properties (5.37) and the orthogonality condition provide the following

$$\alpha_k = \frac{\langle s^\star - s_k, d_k \rangle}{\langle d_k, d_k \rangle} \text{ and } \beta_k = -\frac{\langle d_{k-1}, z_k \rangle}{\langle d_{k-1}, d_{k-1} \rangle}.$$

For $k = 0$, we prescribe $s_0 = 0$ and $d_0 = A(s^\star - s_0)$. The following proposition is a generalisation of [16, Sect. 3].

Proposition 5.7 *If the operator A has the properties*

1. $\langle s, As \rangle > 0$ *for* $s \in \mathscr{S}$ *with* $s \neq 0$,
2. $\langle t, As \rangle = \langle At, s \rangle$ *for* $s, t \in \mathscr{S}$,

then the Krylov subspace method converges.

Construction of A

Let ψ_j, $j = 1, \ldots, 3N$ be a basis of $\mathscr{S}$. The application of A to an element $s \in \mathscr{S}$ is given by

$$As = \sum_{j=1}^{3N} \frac{\langle \psi_j, s \rangle}{\langle \psi_j, \psi_j \rangle} \psi_j.$$

Lemma 5.13 *The operator A fulfils the requirements of the Proposition 5.7.*

Proof Let $t, s \in \mathscr{S}$. It follows, that

$$\langle t, As \rangle = \sum_{j=1}^{3N} \frac{\langle \psi_j, s \rangle \langle \psi_j, t \rangle}{\langle \psi_j, \psi_j \rangle} = \langle At, s \rangle.$$

Furthermore, if we assume $\langle s, As \rangle = 0$, we obtain

$$\langle s, As \rangle = \sum_{j=1}^{3N} \frac{\langle \psi_j, s \rangle^2}{\langle \psi_j, \psi_j \rangle} = 0.$$

Consequently $\langle \psi_j, s \rangle = 0$ for all $j = 1, \ldots, 3N$. Since ψ_j is a basis of $\mathscr{S}$, we can represent s in the form

$$s = \sum_{j=1}^{3N} \theta_j \psi_j.$$

Hence $s = 0$, because

$$\|s\|_{\mathscr{S}}^2 = \langle s, s \rangle = \sum_{j=1}^{3N} \theta_j \langle \psi_j, s \rangle = 0.$$

$\square$

The optimal choice for ψ_j can be described as follows. By the use of the bilinear form to construct orthogonal functions, we have $A\psi_j = \psi_j$ for all $j = 1, \ldots, 3N$. The search direction d_1 is then defined by $d_1 = \alpha_1 s^*$ where $\alpha_1 = 1$. If we choose ψ_j as the Lagrange basis, namely

$$\psi_j = \sum_{l=1}^{N} \Phi(x - x_l) \lambda_{jl}, \ \lambda_{jl} \in \mathbb{R}^3$$

and

$$\psi_j(x_i) = \delta_{\lfloor \frac{i}{3} \rfloor, i} e_k, \quad k = j \bmod 3 + 1, \quad i = 1, \ldots, N,$$

where $e_k \in \mathbb{R}^3$ is the k-th canonical basis vector, it follows that $A\psi_j = \psi_j$ and $\langle \psi_i, \psi_j \rangle = 0$ for $i \neq j$. Unfortunately, the numerical effort for this choice is too high.

Consequently, we have to choose a modified basis

$$\psi_j = \sum_{l = \lfloor \frac{i}{3} \rfloor}^{N} \Phi(x - x_l) \lambda_{jl}, \quad \lambda_{jl} \in \mathbb{R}^3 \tag{5.40}$$

and

$$\psi_j(x_i) = \delta_{\lfloor \frac{i}{3} \rfloor, i} e_k, \quad k = j \bmod 3 + 1, \quad i = \left\lfloor \frac{j}{3} \right\rfloor + 1, \ldots, N.$$

We have

$$\langle \psi_k, \psi_m \rangle = \sum_{i=1}^{N} \lambda_{ki}^\top \sum_{j=1}^{N} \Phi(x_i - x_j) \lambda_{mj} = \sum_{i=1}^{N} \lambda_{ki}^\top \psi_m(x_i)$$

$$= \sum_{j=1}^{N} \psi_k(x_j)^\top \lambda_{mj},$$

where $\lambda_{mj} = 0$ respectively $\lambda_{ki} = 0$ if not listed in (5.40). Hence, we have

$$\langle \psi_k, \psi_m \rangle = \delta_{km}.$$

However, the numerical effort is comparable to that of the full Lagrange basis, so a further modification is necessary. Powell originated the idea of the approximate Lagrange basis [16]. Instead of using the full index set for the summation index l in equation (5.40), only subsets

$$\mathrm{In}_j \subset \{\left\lfloor \frac{j}{3} \right\rfloor, \ldots, N\}$$

of the full index are used. A good choice are the q nearest neighbours of the point x_j. If $N - j + 1 \leq q$ we choose the remaining indices.

Lemma 5.14 *The approximate Lagrange basis is a basis of $\mathcal{S}$.*

Proof Let $C \in \mathbb{R}^{3N \times 3N}$ and write the coefficients λ_{jl} in columns. C is an upper triangular matrix with positive diagonal entries. This follows by

$$(\lambda_{kk})_{k \bmod 3+1} = \sum_{j=1}^{N} \psi_k(x_j)^{\mathsf{T}} \lambda_{kj} = \langle \psi_k, \psi_k \rangle > 0 \text{ for all } k = 1, \ldots, 3N.$$

The last inequality is strict due to $\psi_k \neq 0$.   □

The operator A has the form

$$As = \sum_{j=1}^{3(N-q)} \frac{\langle \psi_j, s \rangle}{\langle \psi_j, \psi_j \rangle} \psi_j + Hs, \text{ for } s \in \mathscr{S},$$

where

$$Hs = \sum_{j=3(N-q)+1}^{3(N-q)} \frac{\langle \psi_j, s \rangle}{\langle \psi_j, \psi_j \rangle} \psi_j.$$

The operator H maps functions $s \in \mathscr{S}$ to the $3q$-dimensional subspace

$$\mathscr{S} \supset \mathscr{S}' = \left\{ s : \sum_{j=N-q+1}^{N} \Phi(x - x_j)c_j, \ c_j \in \mathbb{R}^3 \right\}.$$

For all $s \in \mathscr{S}'$ it follows $Hs = s$. We deduce from

$$\langle \psi_k, s \rangle = \sum_{i=1}^{N} \lambda_{ki}^{\mathsf{T}} s(x_i)$$

that the dependence of Hs on s is only through the values $s(x_i)$, $i = N - q + 1, \ldots, N$. Consequently we have $Hs = Ht$ for the unique element $t \in \mathscr{S}'$ which fulfils the conditions

$$t(x_i) = s(x_i) \text{ for all } i = N - q + 1, \ldots, N.$$

Hs is the solution of the interpolation problem in $\mathscr{S}'$ for $s \in \mathscr{S}$.

Lemma 5.15 *The operator A fulfils the requirements of the Proposition 5.7.*

Algorithm

Input parameters: matrix-valued RBF, point set $X = \{x_i\}_{i=1}^{N}$, size of index set q and the right hand side.

1. Construction of the index sets In_j. Note that two index sets In_j and In_k are identical if $\lfloor \frac{j}{3} \rfloor = \lfloor \frac{k}{3} \rfloor$ holds.
2. Solve the interpolation problems

$$\psi_j(x_i) = \delta_{\left\lfloor \frac{j}{3} \right\rfloor, i} e_k, \quad k = j \bmod 3 + 1, \quad i = 1, \ldots, N.$$

3. Factorise the interpolation Matrix H.

The coefficient vector for a function $s \in \mathscr{S}$ is defined by $\underline{c}(s) \in \mathbb{R}^{3N}$. In the k-th iteration the residuum $r_{k,i} \in \mathbb{R}^3$ is given by

$$r_{k,i} = f_i - s_k(x_i)$$

for $i = 1, \ldots, N$. For $k = 1$ define $r_{1,i} = f_i$. The application of A to $s^\star - s_k$ provides

$$\underline{c}(d_k) = \underline{c}(A(s^\star - s_k)) + \beta_k \underline{c}(d_{k-1}), \quad k \geq 2,$$

where for $k = 1$ the search direction d_{k-1} is omitted. From the orthogonality condition and bilinear form on $\mathscr{S}$, we deduce

$$\beta_k = -\frac{\sum_{i=1}^{N} \underline{c}(A(s^\star - s_k))_i^\top d_{k-1}(x_i)}{\sum_{i=1}^{N} \underline{c}(d_{k-1}) d_{k-1}(x_i)}$$

together with

$$\alpha_k = \frac{\sum_{i=1}^{N} \underline{c}(d_k)_i^\top r_{k,i}}{\sum_{i=1}^{N} \underline{c}(d_k)_i^\top d_k(x_i)}.$$

The coefficients $\underline{c}(s)_i$ consist of the $3i$, $3i + 1$ and $3i + 2$ components of $\underline{c}(s)$.

5.5 Numerical Examples

5.5.1 Krylov Subspace Method

To test the efficiency of the proposed Krylov subspace method, take uniformly distributed (step size is h) points in $[0, 1]^3$. The matrix-valued RBF Φ_{curl} generated by an inverse multiquadric is used with a fixed scaling parameter. The number of Krylov iterations depending on various parameters of the algorithm is shown in the Figs. 5.2 and 5.3. The method is clearly superior to the method of conjugate gradients, which failed to converge in some cases (see Table 5.1). However, one observes that, as the condition number increases (due to the problem size or a change in the scaling parameter), an increase of the parameter q is necessary.

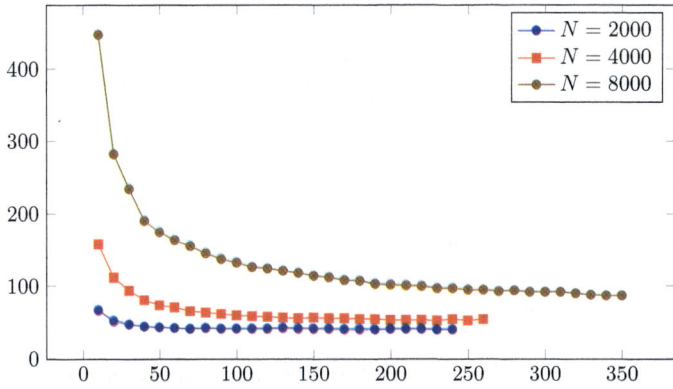

Fig. 5.2 Krylov iterations in dependence of the parameter q

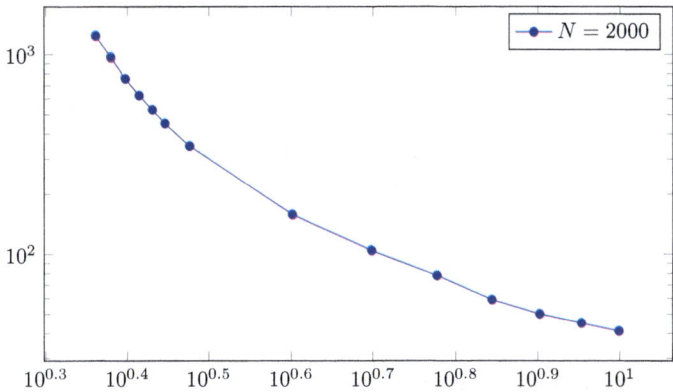

Fig. 5.3 Krylov iterations in dependence of the scaling parameter

Table 5.1 Comparing the CG and the proposed method

N	CG	Krylov ($q = 100$)
1000	170	29
2000	542	41
3000	954	49
4000	1308	60
5000	$>3N$	70

5.5.2 Error Behaviour

We consider the interpolation by means of the matrix valued RBF $\Phi_{\mathrm{Lamé}}$. According to our estimates, the actual error behaviour is limited by both, the smoothness of the right hand side and the RBF ϕ used to generate $\Phi_{\mathrm{Lamé}}$. To generate the right hand side, we use the function introduced by Franke

$$\tilde{f}(x_1, x_2) =$$

$$\frac{3}{4} \exp\left(-\frac{(9x_1 - 2)^2}{4} - \frac{(9x_2 - 2)^2}{4}\right) + \frac{3}{4} \exp\left(-\frac{(9x_1 + 1)^2}{49} + \frac{(9x_2 + 1)}{10}\right)$$

$$+ \frac{1}{2} \exp\left(-\frac{(9x_1 - 7)^2}{4} - \frac{(9x_2 - 3)}{4}\right) + \frac{1}{5} \exp\left(-(9x_1 - 2)^2 - (9x_2 - 7)^2\right),$$

and define the right hand side as

$$f(x_1, x_2, x_3) = (\tilde{f}(x_1, x_2), \tilde{f}(x_2, x_3), \tilde{f}(x_1, x_3))^\top.$$

This function is of the class $C^\infty([0, 1])^3$, thus the interpolation error is dominated by the smoothness of ϕ. To stay in the correct space for the interpolation error estimates, Wendland's functions $\phi_{3,k}$ are used. They are defined by

$$\tilde{\phi}_{3,k}(r) = \mathscr{I}^k \tilde{\phi}_{\lfloor 3/2+k+1 \rfloor}(r), \quad k \in \mathbb{N},$$

where

$$\mathscr{I}\tilde{\phi}(r) = \begin{cases} \int_r^\infty s\tilde{\phi}(s)ds & , r > 0 \\ \mathscr{I}\tilde{\phi}(-r) & , r < 0 \end{cases}.$$

and

$$\tilde{\phi}_l(r) = (1 - r)^l_+, \quad l \in \mathbb{N}.$$

These functions fulfil the algebraic decay condition

$$c_1(1 + \|\xi\|_2)^{-3-2k-1} \le \hat{\phi}_{3,k} \le c_2(1 + \|\xi\|_2)^{-3-2k-1}.$$

For our tests, we use the family

$$\phi_{3,1} = (1 - r)^4_+(4r + 1),$$
$$\phi_{3,2} = (1 - r)^6_+(35r^2 + 18r + 3).$$

The results are log-log plotted in Fig. 5.4. We observe the predicted order of convergence and its dependence on the smoothness of the kernel function.

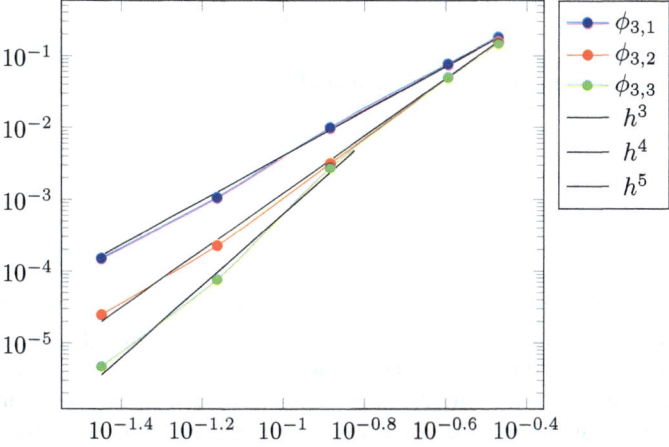

Fig. 5.4 L_2-error in dependence of the step size h

5.5.3 Particular Solution for a Volume Force Density

Consider the Lamé equations

$$-\sum_{j=1}^{3} \frac{\partial}{\partial x_j} \sigma_{ij}(u, x) = f_i(x), \quad x \in \Omega, \ i = 1, 2, 3.$$

The right hand side f has the physical meaning of a density of a volume force. Typical examples are gravitational, electromagnetic, centrifugal, Coriolis and thermal forces. In the following two subsections we illustrate and test our approach on the cases of the gravitation and the centrifugal forces.

Gravitation

The gravitational force acts due to the mass of the body and plays an important role in the static design of bridges and buildings. It has the density

$$f(x) = -\rho g e_3$$

and $e_3 = (0, 0, 1)^\top$, where ρ is the mass density of the material. A particular solution to the Lamé system in this case can be written as

$$u_p(x) = \left(0, 0, \frac{\rho g x_3^2}{2(\lambda + 2\mu)}\right)^\top.$$

We take Ω to be the half sphere

$$\Omega = \{x \in \mathbb{R}^3 : x_3 > 0, \ \|x\|_2 < 5\},$$

and prescribe the Dirichlet data on the boundary

$$\Gamma_D = \{x \in \Omega : x_3 = 0, \ \|x\|_2 < 5\}$$

by

$$(\gamma_0 u)(x) = (0, 0, 0)^\top.$$

The remaining part of the surface is stress free. A quasi-uniform sequence of tetrahedron grids is used for the computation of the error. In Fig. 5.5, the surface tractions for the second refinement are shown.

For the projection of the Dirichlet and Neumann data, a quadrature of order 8 was used. It is important to note that the evaluation matrix can also be approximated by an hierarchical matrix with a good compression ratio and an acceptable accuracy. The parameters of the grids and the matrix compression ratios are listed in Table 5.2.

Fig. 5.5   Surface tractions. Dirichlet (left) and the Neumann (right) boundaries

Table 5.2 Parameters of the grid sequence

Number	Vertices	Elements	Vertices	Elements
	Volume		Surface	
0	380	1368	261	518
1	2386	10,944	1038	2072
2	16,751	87,522	4146	8288
3	125,197	700,416	16,578	33,152
Number	Qaud. points	Compression		
		Quadrature	Interpol.	BEM
0	33,152	0.119	0.81	1.0
1	132,608	0.023	0.50	0.365
2	530,432	0.0158	0.27	0.205
3	2,121,728	0.002	–	0.057

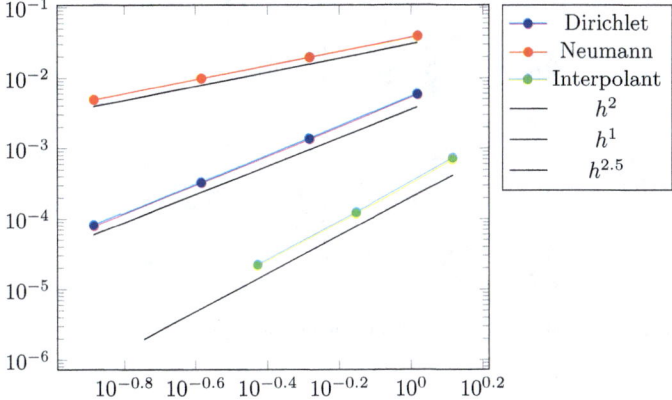

Fig. 5.6 Convergence of the numerical solution in L_2-norm

The BEM, RBF interpolation and RBF evaluation matrices have been approximated with the accuracy 10^{-6}, 10^{-8} and 10^{-10} respectively. To obtain a reference solution, the boundary values have been corrected using the exact particular solution and the homogeneous BVP has been solved by BEM. The relative errors of the Dirichlet and the Neumann data are shown in Fig. 5.6. We observe the expected rates of convergence of the BEM. This implies, that the error in computing the particular solution does not interfere with the overall procedure.

5.5.4 Centrifugal Force

The centrifugal force occurs in rotating bodies. It acts radially outwards from the rotational axis. For fast rotations, e.g. in the case of a crankshaft, the arising force causes a slight deformation of the component. The force has the density

$$f(x) = m\omega^2 r,$$

where r is the radius vector to the axis of rotation. In our example, a crankshaft, rotating about the x_1-axis is considered. The boundary values are given by

$$(\gamma_0 u)(x) = 0,$$

for

$$x \in \Gamma_D = \{x \in \Gamma : \ x_1 = -2 \text{ or } x_1 = 8.25\}.$$

The remaining part of the boundary is stress free. In Fig. 5.7 the deformed crankshaft is shown. The arrows depict the displacement vector, the surface colour shows the

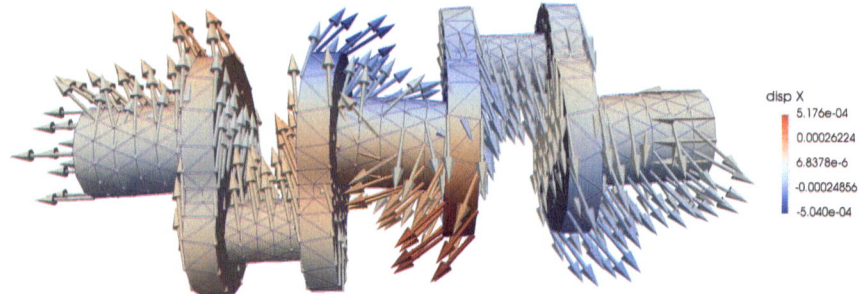

disp X
5.176e-04

0.00026224

6.8378e-6

-0.00024856

-5.040e-04

Fig. 5.7 Deformed crankshaft (displacement scaling factor 200)

displacement in x_1 direction. The values $E = 75$ GPa and $v = 0.2$ were used for the material parameters.

5.6 Application to Fibre Reinforced Plastic

5.6.1 Linear Case

If the BVP is linear and homogeneous, our method reduces to H^2-matrix accelerated BEM. We demonstrate the efficiency of the method on geometries relevant to modelling composite materials on the level of micro structure. Consider a series of Representative Volume Elements (RVEs) consisting of a soft matrix material with fibre-like inclusions (see Fig. 5.8). The inclusions are randomly oriented and their

Fig. 5.8 RVE with uniform distributed inclusions

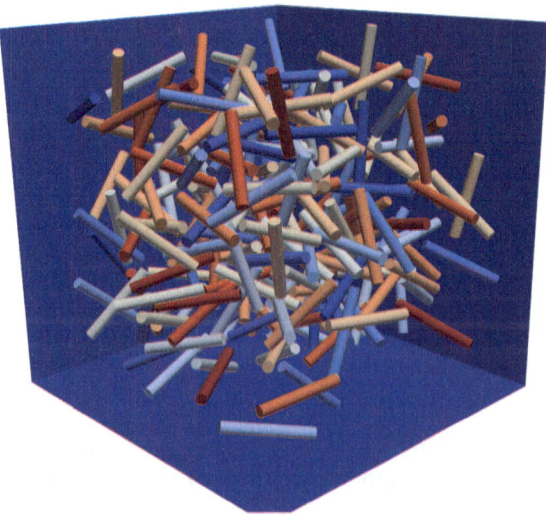

Table 5.3 Computing time for the homogenisation

	227	150	100	50	25	0
Triangles	51,458	34,344	23,204	12,088	6534	978
Vertices	26,185	17,474	11,804	6146	3319	491
Time in h	2.9	1.06	0.47	0.15	0.08	–

Table 5.4 Compression of the potentials, 227 inclusions, accuracy 10^{-6}

Matrix	K	V_0	V_1	V_2	V_3	V_4	V_5	V_6
Compression (%)	11.7	4.1	6.8	7.3	7.3	6.9	7.3	6.9

number varies from 0 to 227 (volume fraction about 2.8%). In this case, the multi-domain formulation from the Sect. 5.2.2 has to be employed. The parameters of the meshes and the solution time for one Dirichlet BVP are listed in Table 5.3. We observe almost linear dependency of the solution time on the number of vertices in the mesh.

The compression ratio of the H^2-matrix of the single and double layer potentials are listed in Table 5.4.

Tensile Test

A simple way to characterise material properties is the tensile test. Consider a thin solid body shaped as in Fig. 5.10. The left and right sides are pulled apart, while the rest of the boundary is stress free. The quantity of interest is the integral reaction force, which dependence on the pulled distance is usually plotted.

On the macro scale we employ a standard Finite Element Method (FEM), but in each its quadrature point the stress strain relation is computed by homogenising a response of a RVE. This homogenisation is performed by the coupled BEM-RBF method.

On the macro level use 224 elements, the relations on the micro scale are evaluated in 8 Gauss quadrature points for every element (see Fig. 5.10) and every macro load step. The homogenisation step is treated by BEM with different RVEs (227 uniformly distributed inclusions, 25 unidirectional inclusions). The unidirectional inclusions are orientated along the horizontal axis. Thus, one expects a larger force response compared to the RVE with uniformly distributed inclusions. In Fig. 5.9 force-displacement curves for different RVE are shown.

Homogeneous Material

As the first non-linear example we consider a RVE without inclusions. The BEM-RBF coupled method is used for homogenisation step on the micro scale. The particular solutions generated by the radial basis functions are relatively small compared to the solution, because the damage variable is nearly constant (up to rounding errors). Hence

$$\operatorname{div}(d\sigma\,(u_B + u_R)) \approx 0.$$

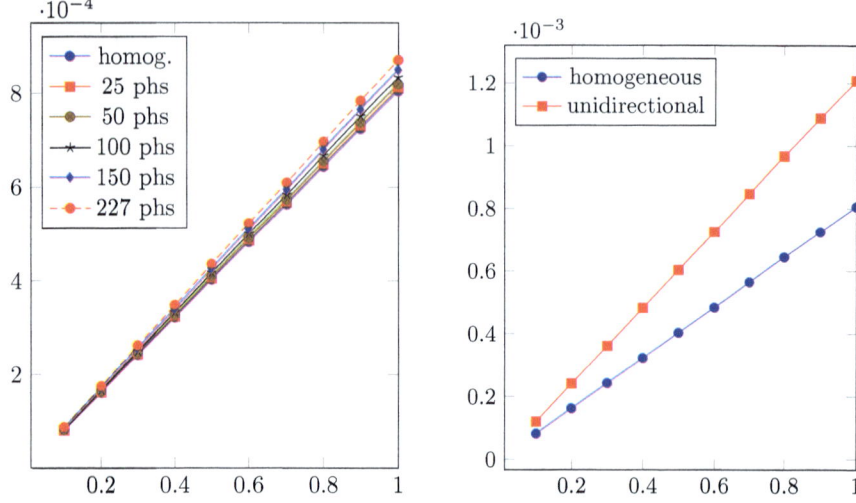

Fig. 5.9 Force (N)-displacement (mm) curves

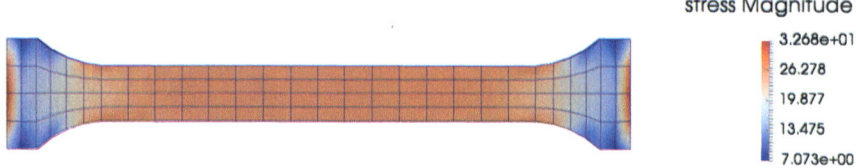

Fig. 5.10 Tensile test. The surface colour shows the surface traction

As a consequence, the fixed point iterations converge after only one step. For the tensile test, on macro-scale again FEM with 8 quadrature points per element is used. In contrast to the linear case, we observe a sub-linear behaviour in the force (N)-displacement (mm) curve (see Fig. 5.11).

5.6.2 RVE with a Spherical Inclusion

We now consider a RVE with a spherical inclusion (Fig. 5.12). The discretisation parameters are as follows

- 312 triangles, 158 vertices on the surface of the sphere,
- 1664 triangles, 834 vertices on the surface of the RVE bounding box,
- 10,979 tetrahedrons in matrix material, with 1339 vertices inside,
- 14,294 interpolation points,
- 126,464 quadrature points for the projection of the interpolation on the surfaces.

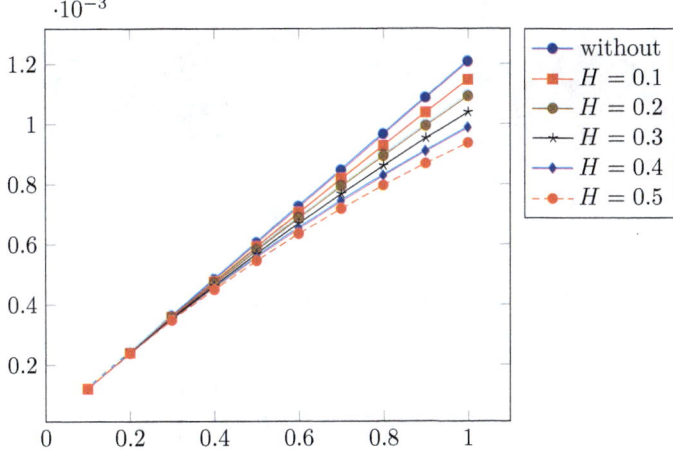

Fig. 5.11 Tensile test for homogeneous non-linear material

Fig. 5.12 RVE with
spherical inclusion

To have a differentiable right hand side $d\sigma(u_R + u_B)$, we are here limited to the case
$Y_0 = 0$. If we omit this restriction, the differentiation of the right hand side leads to
oscillations in the stress tensor, and therefore a higher energy. The parameter H is
critical. In a first experiment H is chosen in such a way, that the damage variable is
less then 0.20. The linear displacement field is prescribed on the outer boundary

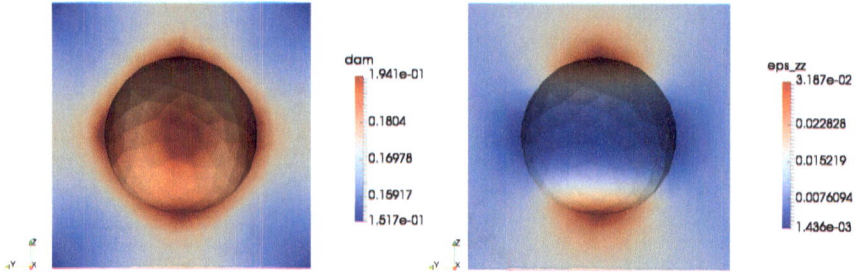

Fig. 5.13 Cut of the RVE with the yz-plane. Damage (left) and e_{zz} (right)

$$u = ex, \ e = \begin{pmatrix} 0.01 & 0 & 0 \\ 0 & 0.01 & 0 \\ 0 & 0 & 0.01 \end{pmatrix}.$$

For the damage model we use the parameters

$$Y_0 = 0, \ H = 0.1.$$

The material parameters are given by

$$E_{\text{mat}} = 2.5\,\text{GPa}, \ E_{\text{phs}} = 75\,\text{GPa}, \ \nu_{\text{mat}} = 0.35, \ \nu_{\text{phs}} = 0.2.$$

In Fig. 5.13 the solution after 13 fixed point iterations is shown. The inclusion was removed from the view, and the values of the damage variable or e_{zz} can be seen. The computational effort per iteration step is rather high due to the projection of the solution of the last step and the particular solution on the boundary. An iteration step took several hours. In Fig. 5.14, the error

$$\frac{\|e_{i+1} - e_i\|}{\|e_i\|}$$

is plotted against the number of steps. The fixed point iteration reaches an accuracy of 10^{-7} after 13 steps. It is a stable fixed point.

To illustrate the effect of the damage variable on the conormal derivative, a line plot was made through the points $(0, 0, 5)$ and $(0, 0, -5)$ (see Figs. 5.15, 5.16, 5.17). It can be seen that, compared to the elastic predictor (step 0), the damage decreases at the boundary, but increases significantly towards the inclusion. The reason for this is that the contact between matrix and inclusion is dictated by the interface conditions in the form of coupled Dirichlet data. In a simplified way, these can also be understood as how much the inclusion must shift in order to generate the corresponding stress. However, the stress is scaled by the factor $(1 - d)$. In the present example, the value of $(1 - d)$ is about 0.8 on the inner surface. This causes a smaller displacement on

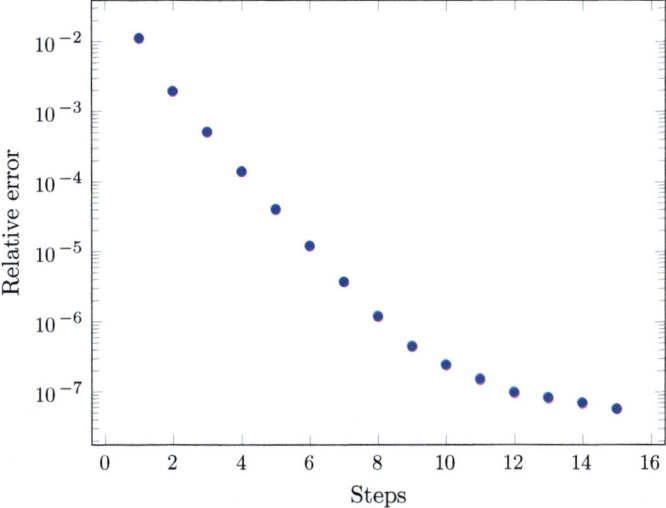

Fig. 5.14 Convergence of the fix point iterations

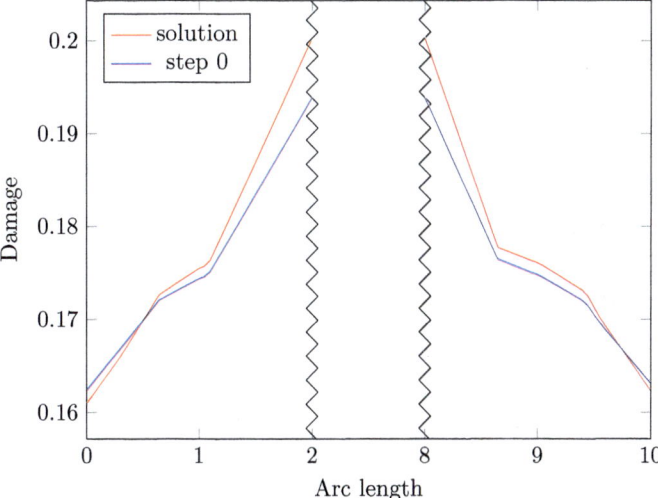

Fig. 5.15 Course of the damage variable

the boundary of the inclusion. In particular, in the regions of damage in the matrix material the strain e increases accordingly.

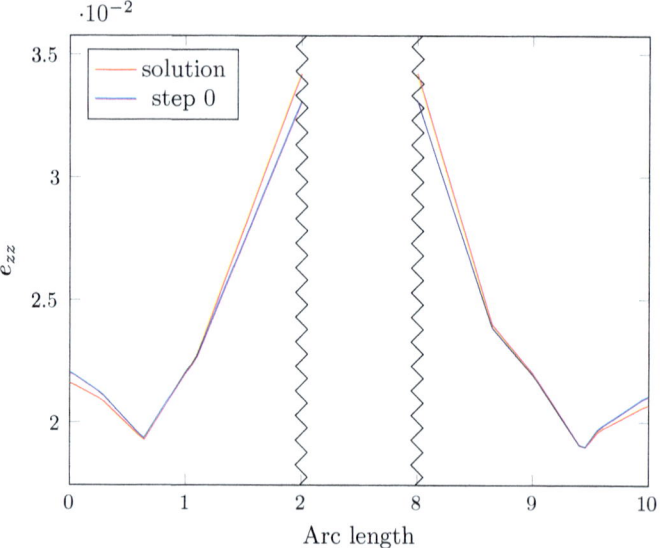

Fig. 5.16 Course of the e_{zz}-component

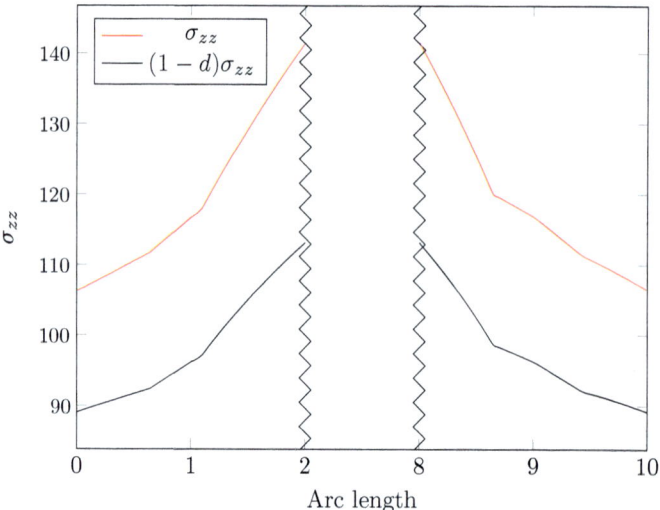

Fig. 5.17 Courses of σ_{zz}-component with and without damage

References

1. Andrä, H., Grzhibovskis, R., Rjasanow, S.: Boundary element method for linear elasticity with conservative body forces. In: Advanced Finite Element Methods and Applications. Lecture Notes Applied Computer Mechanics, vol. 66, pp. 275–297. Springer, Heidelberg (2013)
2. Arad, N.: Image warping by radial basis functions: applications to facial expressions. CVGIP Graph. Models Image Process. **56**(2), 161–172 (1994)
3. Beatson, R.K., Light, W.A.: Fast evaluation of radial basis functions: methods for two-dimensional polyharmonic splines. IMA J. Numer. Anal. **17**(3), 343–372 (1997)
4. Beatson, R.K., Newsam, G.N.: Fast evaluation of radial basis functions. Advances in the theory and applications of radial basis functions. I. Comput. Math. Appl. **24**(12), 7–19 (1992)
5. Beatson, R.K., Cherrie, J.B., Mouat, C.T.: Fast fitting of radial basis functions: methods based on preconditioned GMRES iteration. Radial basis functions and their applications. Adv. Comput. Math. **11**(2-3), 253–270 (1999)
6. Bebendorf, M.: Approximation of boundary element matrices. Numer. Math. **86**(4), 565–589 (2000)
7. Bebendorf, M.: Hierarchical matrices. A means to efficiently solve elliptic boundary value problems. In: Lecture Notes in Computational Science and Engineering, vol. 63. Springer, Berlin (2008)
8. Bebendorf, M., Grzhibovskis, R.: Accelerating Galerkin BEM for linear elasticity using adaptive cross approximation. Math. Meth. Appl. Sci. **29**, 1721–1747 (2006)
9. Bebendorf, M., Rjasanow, S.: Adaptive low-rank approximation of collocation matrices. Computing **70**, 1–24 (2003)
10. Börm, S.: Efficient numerical methods for non-local operators. H^2-matrix compression, algorithms and analysis. In: EMS Tracts in Mathematics, vol. 14. European Mathematical Society (EMS), Zürich (2010)
11. Buhmann, M.D.: Radial basis functions: theory and implementations. In: Cambridge Monographs on Applied and Computational Mathematics, vol. 12. Cambridge University Press, Cambridge (2003)
12. Cheng, A.: Particular solutions of Laplacian, Helmholtz-type, and polyharmonic operators involving higher order radial basis function. Eng. Anal. Boundary Elem. **24**, 531–538 (2000)
13. Cherrie, J.B., Beatson, R.K., Newsam, G.N.: Fast evaluation of radial basis functions: methods for generalized multiquadrics in $\mathbb{R}^n$. SIAM J. Sci. Comput. **23**(5), 1549–1571 (2002)
14. Cortes, C., Vapnik, V.: Support-vector networks. Mach. Learn. **20**, 273–297 (1995), https://doi.org/10.1023/A:1022627411411
15. Fasshauer G.E.: Meshfree approximation methods with MATLAB. In: Interdisciplinary Mathematical Sciences, vol. 6. World Scientific Publishing Co., Pte. Ltd., Hackensack, NJ: With 1 CD-ROM. Windows, Macintosh and UNIX (2007)
16. Faul, A.C., Powell, M.J.D.: Krylov subspace methods for radial basis function interpolation. In: Numerical Analysis 1999 (Dundee), Chapman & Hall/CRC Res. Notes Math., vol. 420, pp. 115–141. Chapman & Hall/CRC, Boca Raton, FL (2000)
17. Fornberg, B., Piret, C.: A stable algorithm for flat radial basis functions on a sphere. SIAM J. Sci. Comput. **30**(1), 60–80 (2007/08)
18. Fuselier, E.J.: Refined error estimates for matrix-valued radial basis functions. Thesis (Ph.D.)— Texas A&M University. ProQuest LLC, Ann Arbor, MI (2006)
19. Fuselier, E.J.: Improved stability estimates and a characterization of the native space for matrix-valued RBFs. Adv. Comput. Math. **29**(3), 311–313 (2008)
20. Fuselier, E.J.: Sobolev-type approximation rates for divergence-free and curl-free RBF interpolants. Math. Comp. **77**(263), 1407–1423 (2008)
21. Giesl, P., Wendland, H.: Meshless collocation: error estimates with application to dynamical systems. SIAM J. Numer. Anal. **45**(4), 1723–1741 (2007)
22. Grzhibovskis, R., Bambach, M., Rjasanow, S., Hirt, G.: Adaptive cross-approximation for surface reconstruction using radial basis functions. J. Eng. Math. **62**(2), 149–160 (2008)

23. Grzhibovskis, R., Michel, C., Rjasanow, S.: Matrix-valued radial basis functions for the Lamé system. Math. Methods Appl. Sci. **41**, 6080–6107 (2018)
24. Hackbusch, W.: A sparse matrix arithmetic based on H-matrices. I. Introduction to H-matrices. Computing **62**(2), 89–108 (1999)
25. Hackbusch, W.: Hierarchische Matrizen: Algorithmen und Analysis. Springer (2009)
26. Hackbusch, W., Khoromskij, B.N.: A sparse H-matrix arithmetic: general complexity estimates. J. Comput. Appl. Math. **125**(1–2), 479–501 (2000); Numerical analysis 2000, Vol. VI, Ordinary differential equations and integral equations
27. Hackbusch, W., Khoromskij, B.N.: A sparse H-matrix arithmetic. II. Application to multidimensional problems. Computing **64**(1), 21–47 (2000)
28. Hackbusch, W., Khoromskij, B.N., Sauter, S.A.: On H^2-matrices. In: Lectures on Applied Mathematics (Munich, 1999), pp. 9–29. Springer, Berlin (2000)
29. Iske, A.: Scattered data modelling using radial basis functions. In: Tutorials on Multiresolution in Geometric Modelling, pp. 205–242. Springer (2002), https://doi.org/10.1007/978-3-662-04388-2_9
30. Iske, A., Le Borne, S., Wende, M.: Hierarchical matrix approximation for kernel-based scattered data interpolation. SIAM J. Sci. Comput. **39**(5), A2287–A2316 (2017)
31. Lee, D.: Fast multiplication of a recursive block Toeplitz matrix by a vector and its application. J. Complex. **2**, 295–305 (1986)
32. Lemaitre, J., Chaboche, J.L.: Aspects phénoménologiques de la rupture par endommagement. Journal de Mecanique Appliquée **2**, 317–365 (1978)
33. Liu, Y., Liew, K., Hon, Y., Zhang, X.: Numerical simulation and analysis of an electroactuated beam using a radial basis function. Smart Mater. Struct. **14**(6), 1163–1171 (2000)
34. Livne, O.E., Wright, G.B.: Fast multilevel evaluation of smooth radial basis function expansions. Electron. Trans. Numer. Anal. **23**, 263–287 (2006)
35. Lowitzsch, S.: A density theorem for matrix-valued radial basis functions. Numer. Alg. **39**(1–3), 253–256 (2005)
36. Lowitzsch, S.: Matrix-valued radial basis functions: stability estimates and applications. Adv. Comput. Math. **23**(3), 299–315 (2005)
37. Muleshkov, A.S., Golberg, M.A.: Particular solutions of the multi-Helmholtz-type equation. Eng. Anal. Boundary Elem. **31**(7), 624–630 (2007)
38. Muleshkov, A.S., Golberg, M.A., Chen, C.S.: Particular solutions of Helmholtz-type operators using higher order polyharmonic splines. Comput. Mech. **23**(5–6), 411–419 (1999)
39. Narcowich, F.J., Ward, J.D.: Norms of inverses for matrices associated with scattered data. In: Curves and Surfaces (Chamonix-Mont-Blanc, 1990), pp. 341–348. Academic Press, Boston, MA (1991)
40. Narcowich, F.J., Ward, J.D.: Generalized Hermite interpolation via matrix-valued conditionally positive definite functions. Math. Comp. **63**(208), 661–687 (1994)
41. Narcowich, F.J., Ward, J.D., Wendland, H.: Sobolev bounds on functions with scattered zeros, with applications to radial basis function surface fitting. Math. Comp. **74**(250), 743–763 (2005)
42. Narcowich, F.J., Ward, J.D., Wendland, H.: Sobolev error estimates and a Bernstein inequality for scattered data interpolation via radial basis functions. Constr. Approx. **24**(2), 175–186 (2006)
43. Powell, M.: Tabulation of thin-plate splines on a very fine two dimensional grid. University of Cambridge, Technical Report No. DAMTP 1992/NA2 (1992)
44. Powell, M.J.D.: Truncated Laurent expansions for the fast evaluation of thin plate splines. Numer. Algorithms **5**(1-4), 99–120 (1993); Algorithms for approximation, III (Oxford, 1992)
45. Rjasanow, S., Steinbach, O.: The Fast Solution of Boundary Integral Equations. No. 12 in Springer Series in Mathematical and Analytical Technology with Applications to Engineering. Springer, Berlin-Heidelberg-New York (2007)
46. Schaback, R.S., Chen, C.S., Hon, R.S.: Scientific computing with radial basis functions. Technical Report, Univeristy of Goettingen (2007)
47. Stein, E.M.: Singular Integrals and Differentiability Properties of Functions, vol. 30. Princeton University Press (1970)

48. Tsai, C.C., Cheng, A.H.D., Chen, C.S.: Particular solutions of splines and monomials for polyharmonic and products of Helmholtz operators. Eng. Anal. Bound. Elem. **33**(4), 514–521 (2009)
49. Uhlir, K., Skala, V.: Radial basis function use for the restoration of damaged images. Computer Vision and Graphics, pp. 839–844. Kluwer Academic Publishers (2006)
50. Wendland, H.: Scattered data approximation. In: Cambridge Monographs on Applied and Computational Mathematics, vol. 17. Cambridge University Press, Cambridge (2005)
51. Wendland, H.: On the stability of meshless symmetric collocation for boundary value problems. BIT **47**(2), 455–468 (2007)
52. Xiang, S., Wang, K.: Free vibration analysis of symmetric laminated composite plates by trigonometric shear deformation theory and inverse multiquadric rbf. Thin-Walled Struct. **47**, 304–310 (2007)
53. Yao, G.: Local radial basis function methods for solving partial differential equations. Ph.D. Thesis. University of Southern Mississippi (2011)
54. Zgurovsky, M.Z., Zaychenko, : Y.P.: Neural networks. In: The Fundamentals of Computational Intelligence: System Approach, pp. 1–38. Springer (2017)

Chapter 6
Experimental Studies

Céline Röhrig and Stefan Diebels

6.1 Available Specimen and Preparation

The material studied in this work is a short fibre-reinforced composite. The matrix consists of polybutylene terephthalate (PBT) which is a semi-crystalline thermoplastic. It is reinforced by short glass fibres (GF, E-type) to increase the stiffness and strength leading to a high geometrical stability in applications. The fibres have a cylindrical geometry. They possess diameters in the range of $d = 10$–$12\,\mu$m and averaged lengths of $l = 200\,\mu$m. The polymeric matrix material PBT is chosen because of its high resistance against outer influences like changes in humidity or contact to chemicals. In order to investigate the influence of different fibre contents different specimens are available consisting either of the pure matrix material or with different fibre contents from 5, 20 and 30 wt% of glass fibres. A materialographic picture of a cross section is shown in Fig. 6.1. The picture presents the cross-section from a directly extruded tensile bar with 5 wt% GF which corresponds to the lowest available fibre content. The different orientations of the glass fibres can be clearly seen.

In addition to the different fibre contents the specimens are available in different geometries. The composite material is either directly extruded to tensile bars which are firstly used in the uniaxial tensile tests or to thin sheets with a cross-section of $150 \times 150\,$mm^2 and a thickness of $2\,$mm. The sheet specimens are used for the following investigations concerning the influence of the fibre orientation and for the multiaxial loading experiments. For this purpose, specimens of different shapes are milled out from these sheets.

C. Röhrig (✉) · S. Diebels
Lehrstuhl für Technische Mechanik, Universität des Saarlandes, Campus A 4.2,
66123 Saarbrücken, Germany
e-mail: celine.roehrig@uni-saarland.de

S. Diebels
e-mail: s.diebels@mx.uni-saarland.de

© Springer-Verlag GmbH Germany, part of Springer Nature 2019
S. Diebels and S. Rjasanow (eds.), *Multi-scale Simulation of Composite Materials*,
Mathematical Engineering, https://doi.org/10.1007/978-3-662-57957-2_6

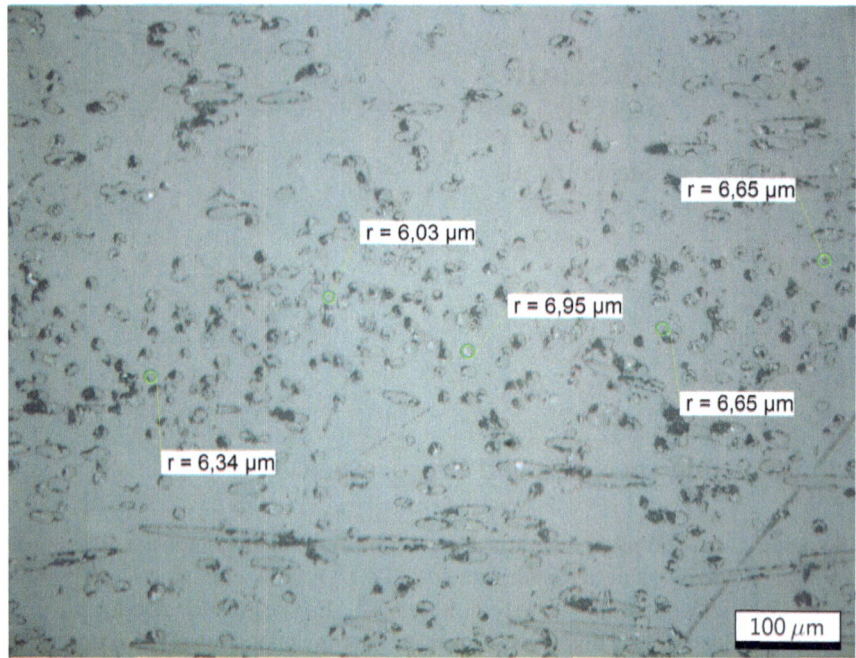

Fig. 6.1 Materialographic specimen: cross-section tensile bar type 1A PBT GF5

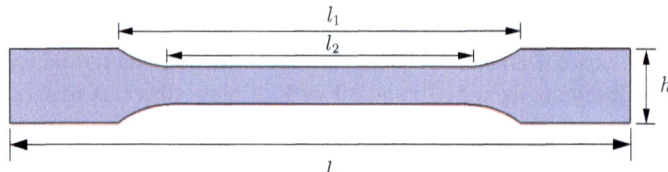

Fig. 6.2 Tensile bar according to DIN EN ISO 527-2 type 1A

6.1.1 Tensile Bars

For the first series of uniaxial tensile tests the directly extruded tensile bars type 1A of DIN EN ISO 527-2 [2] are used. Figure 6.2 presents the geometry of these bars of type 1A. The dimensions are $l = 170$ mm, $l_1 = 110$ mm, $l_2 = 80$ mm, $h = 20$ mm and $d = 4$ mm.

Using these bars, uniaxial tests at different strain rates and cyclic loading tests are carried out in order to classify the material behaviour of the matrix material and the composites with different fibre contents. For a detailed investigation of the fibre orientation smaller tensile specimens are milled out of the sheet material.

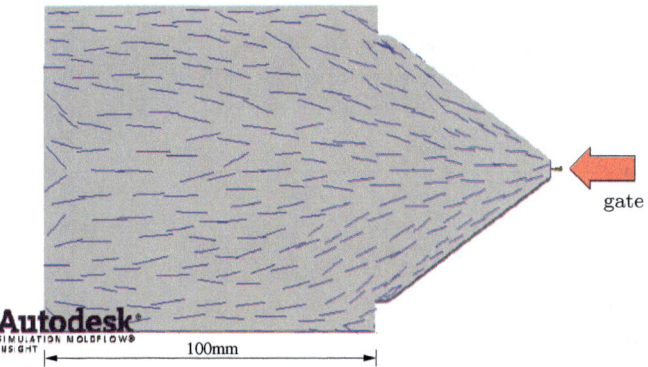

Fig. 6.3 Moldflow®-simulation: process-induced fibre orientation

Fig. 6.4 Specimen
preparation by milling

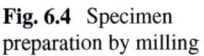

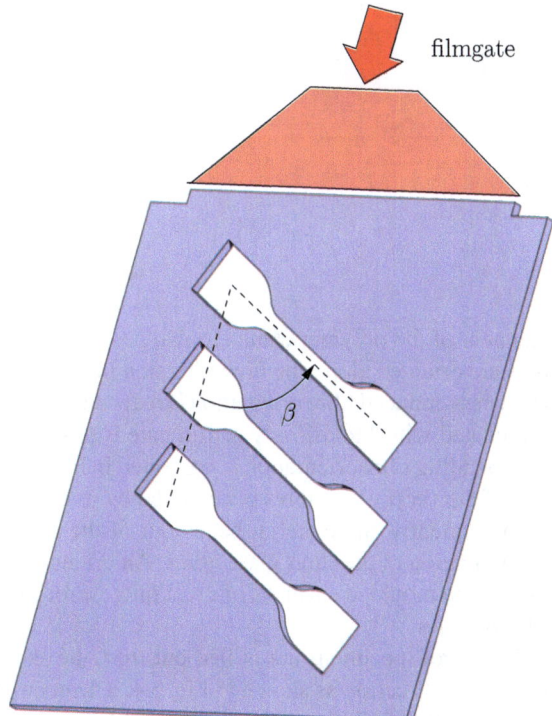

6.1.2 Sheet Material

Based on the production process of injection moulding of the polymeric composite a certain fibre orientation is achieved depending on the geometry of the cavity, where the flow gradient of the viscous polymer melt influences an alignment of the short fibres. The process-induced fibre alignment can be seen in Fig. 6.1. In addition, an

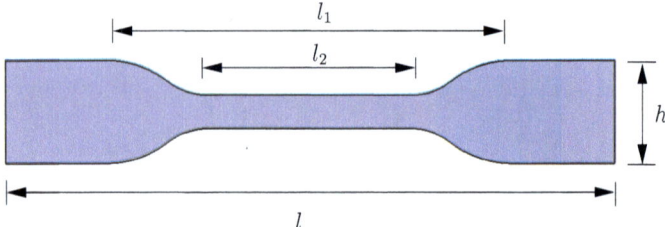

Fig. 6.5 Tensile bar according to DIN EN ISO 527-2 type 5A

Fig. 6.6 Tensile bars with
fibre orientations β as
expected to the main flow
direction during the
moulding of the sheets

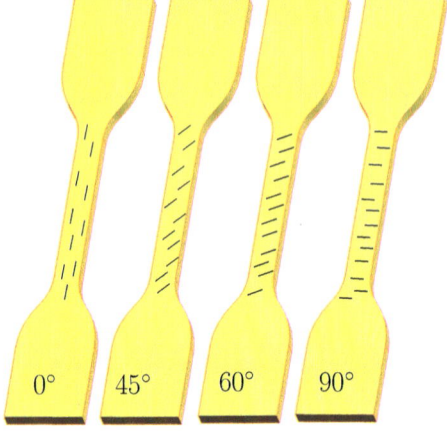

alignment of the polymer chains is also possible. Therefore, most of the short glass
fibres are oriented along the flow direction in the process. A further visualisation of
the process-induced fibre orientation is realised. The filling process of the plate cavity
is simulated with Moldflow®. A film gate is positioned on the right hand side. The
corresponding fibre orientation is presented in Fig. 6.3. As a basis the assumed alignment of the short glass fibres is approved by microsections and simulation results.

Consequently, the material behaviour of the composite is mainly influenced by
this orientation of the short glass fibres. An orientation distribution of the short glass
fibres in the moulded sheets is of great interest for the investigations of the mechanical properties.

Therefore, specimens are milled out from the sheets in different angles β to the
main filling direction as shown in Fig. 6.4. The geometry of these small tensile bars
of DIN EN ISO 527-2 type 5A is shown in Fig. 6.5. The dimensions are $l = 75$ mm,
$l_1 = 50$ mm, $l_2 = 25$ mm, $h = 12.5$ mm and $d = 2$ mm.

With this preparation procedure, small tensile bars with different main fibre orientations are extracted from the sheets. For the uniaxial tensile tests, samples with
main fibre orientations in parallel to the tensile direction (0°), rectangular to the tensile direction (90°) and two different directions in between (45°, 60°) are available.
Figure 6.6 shows the positions of the different samples with respect to the main flow
direction during filling the cavity.

6.2   Digital Image Correlation and Strain Measurement

In the next section, results of uniaxial tensile tests are shown. The first series of experiments is used to characterise the general material behaviour of the PBT matrix as well as the different composites. The classical evaluation of the uniaxial tests is based on the assumption of homogeneous distributions of stresses and strains. Therefore, the stress in axial direction of the specimens is computed by the ratio of the measured force and the initial cross-sectional area of the sample while the corresponding strain follows by the displacement of the device's clamps and the initial length of the sample. This simple evaluation of the experiments can only be used as a first guess. A detailed investigation shows that the assumption of homogeneity is not valid if large deformations and damage occur. In this case, localisation takes place and a necking region forms. In this area the local deformation is much larger than the average strain and the cross-section narrows down. These effects need to be resolved by a local measurement. Touching methods will not lead to a satisfying solution of the problem. On the one hand, strain gauges applied to the specimen's surface may affect the measurement itself by the high stiffness of the glue, which is needed during the application. On the other hand, strain gauges allow for an local measurement only at the position where they are applied. They do not deliver a full field measurement. But this is necessary because the region of the localisation zone is arbitrary and not known a priori. Therefore, a contactless measurement system is chosen for the evaluation of the experimental data concerning the specimen deformation. As a further advantage, this system will provide a full field measurementand can also be used for the evaluation of the following multiaxial experiments. In experiments, like true biaxial experiments or the Nakajima test, the strain is inhomogeneously distributed from the very beginning. Therefore, a full field measurement of the local deformation is indispensable. Hence, using an optical deformation measurement system to evaluate the experimental data over the entire specimen surface is the chosen method. In this procedure a series of pictures of the deforming specimen are taken during the whole experiment. The recorded pictures of the specimen's surface are then evaluated using a digital image correlation (DIC) software [3, 4, 6, 17, 22]. The DIC software requires a stochastic pattern in each picture and tries to correlate the patterns in the deformed and undeformed case. If the examined material has a visible microstructure like foams [12, 13] or linen, a special specimen preparation before experimental testing is not required. But if there is no stochastic microstructure visible on the surface, a speckle pattern has to be applied, e.g. a point cloud of varnish particles is sprayed on the specimen surface e.g. by airbrush techniques. In both cases, the DIC is based on a comparison of the gray values in the structures of the images taken during the experiment.

In the following investigations a stochastic speckle pattern is applied onto the surface of the composite as shown in Fig. 6.7.

The DIC software divides the pixels of an image in so-called subsets, e.g. small parts of the whole image, and evaluates the gray values in the picture sequences. During the deformation the software compares the gray values and their distribution

Fig. 6.7 Speckled tensile bar type 1A

in the subsets with the undeformed state. The comparison is evaluated based on the cross-correlation coefficient C_n according to Eq. (6.1)

$$C_n = \sum_i \sum_j \frac{(G_0(x_i, y_i) - \overline{G}_0)(G_t(x_i', y_i') - \overline{G}_t)}{\sqrt{\left(\sum_i \sum_j (G_0(x_i, y_i) - \overline{G}_0)^2\right)\left(\sum_i \sum_j (G_t(x_i', y_i') - \overline{G}_t)^2\right)}} .$$

(6.1)

In this relation, $G_0(x_i, y_i)$ represents the gray value of the pixel i located in position x_i and y_i in the reference picture and $\overline{G}_0$ is the mean value of the considered subset. $G_t(x_i', y_i')$ and $\overline{G}_t$ are the corresponding gray values in the deformed state at time t. The mean values $\overline{G}_0$ and $\overline{G}_t$ are substracted in order to eliminate local differences in the brightness. The coordinates of a pixel in the reference picture are (x_i, y_i), whereby the same identified pixel in the picture of the deformed state is located at position (x_i', y_i'). The pixel coordinates define the displacement and the deformation of the subsets. According to Eqs. (6.2) and (6.3), u and v are the displacements of the pixels in the two perpendicular directions e_1, e_2 while Δx and Δy are the positions of the observed pixels in the masked area

$$x' = x + u + \frac{\partial u}{\partial x} \Delta x + \frac{\partial u}{\partial y} \Delta y = x + d_x , \tag{6.2}$$

$$y' = y + v + \frac{\partial v}{\partial x} \Delta x + \frac{\partial v}{\partial y} \Delta y = y + d_y . \tag{6.3}$$

For the correlation procedure, the masked area is divided into small subsets, which are deformed, displaced and rectified until the gray values of the deformed subset are similar to the reference picture, cf. Fig. 6.8 [19]. The deformation field could be calculated from the displacements of the pixels in the subsets by means of an optimisation process considering the basics of continuum mechanics. The Lucas–Kanade-algorithm Eq. (6.4) [14] is commonly used to solve the optimisation problem

$$arg \underset{d}{min}\left(\sum_{i,j} ||G_i(x + d_x, y + d_y) - G(x, y)||\right). \tag{6.4}$$

The in-plane strain is obtained as full-field information if all pixels are taken into account which lay in the masked area of the specimen's surface. The out-of-plane

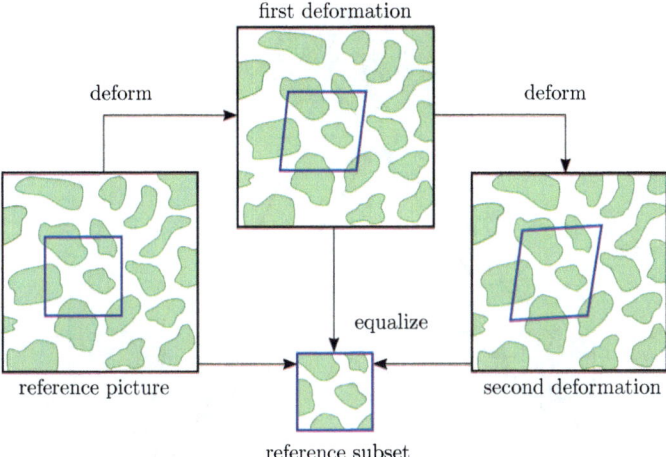

Fig. 6.8 Scheme of digital image correlation

displacement cannot be obtained because two different points (X, Y) may be projected to the same point $P(X, Y)$ in the image plane of the camera as sketched in Fig. 6.9. But using a calibrated setup with two cameras also allows to compute the out-of-plane displacement based on a stereoscopic evaluation of the images of both cameras. Even if two cameras are used a thickness change of the specimens cannot be detected because the measured out-of-plane displacement may result from a rigid body motion of the specimens. Therefore, two systems each consisting of two cameras have to be placed in front and rear of the specimens. This setup can be used to compute the thickness change of the specimens by taking the difference between the out-of-plane displacement on the front and rear sides. If homogeneity over the thickness d of the sample is assumed the normal strain in the third direction $\mathbf{e_3}$ is obtained from this information.

The influence of compressibility or anisotropic effects is of great interest concerning the experimental characterisation of a thermoplastic composite material. For this purpose the displacement information in all three spatial directions (see Fig. 6.10) is needed in the analysed volume element during the experiment. Using a four camera setup as shown in Fig. 6.11 allows not only to measure the in-plane strains on the front and the rear of the specimens but also to measure the deformation in the thickness direction during the strain localisation in the necking area.

Following this method, four pictures, which are triggered at the same time, have to be taken by the four cameras during each load step of the experiment. Before starting any measurement, the commercial digital image correlation software *ISTRA 4D®* provided by Dantec Dynamics® [8], which is used in the following to calculate all strains and displacements, has to know the exact adjustment of the four cameras. Therefore a calibration procedure is needed. For this purpose a calibration target with an exactly defined thickness and an imprinted coordinate system is photographed by

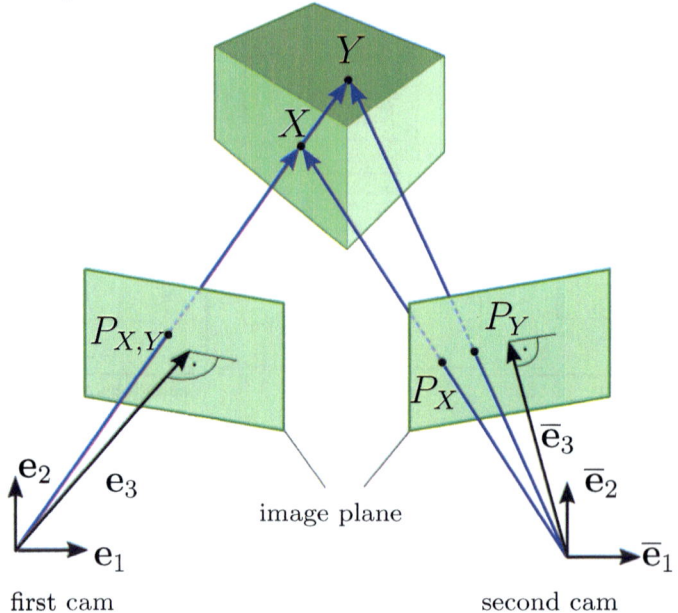

Fig. 6.9 Scheme of stereoscopy

Fig. 6.10 3D-deformation
of a cube

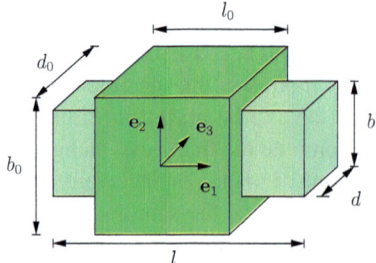

Fig. 6.11 Four camera setup
with a specimen in the centre

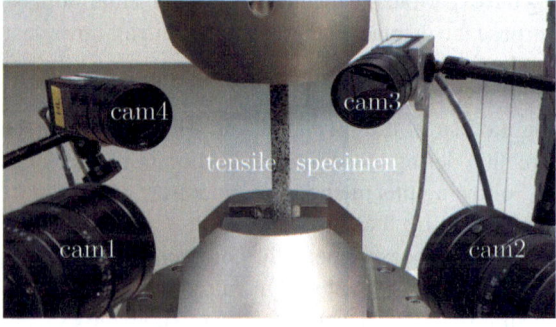

camera 1   camera 2   camera 3   camera 4

Fig. 6.12 Four camera setup: calibration photos and two stadiums of specimen in experiment (1) initial step, (2) deformed state [18]

all cameras. Several photos are taken from the target in different positions. By the exact knowledge of the target's geometry the relative positions of the cameras are automatically obtained by the software *ISTRA 4D®*. Finally these positions allow for the calibration of all cameras to each other. The presented system has the possibility of an expansion with up to 16 cameras in order to measure the full-field strains in a complex geometry by means of a 360°-3D-optical evaluation.

Completing a calibration procedure the correlation software is able to evaluate the experimental data by comparing successively recorded photos with the undeformed shape of the specimen as reference. The special feature of the used 360°-3D-measurement system with the DIC-software *ISTRA 4D®* [8] is the calculation of the two in-plane strains and additionally of the out-of-plane strain of the specimen. In this case, the evaluation of all spatial strains on the material surface is guaranteed because the described measurement system realises the perfect superposition of the information from the front side and the rear side of the tensile specimen. The data acquisition for the four camera setup of the uniaxial tested tensile bar is represented in Fig. 6.12. The upper row of pictures shows the calibration measurement, which is important for the identification of the exact camera positions to each other. The second line represents the initial state of the specimen and the last line shows the deformed specimen during uniaxial tension including a necking area. While the required speckle pattern can be clearly seen on all images in the undeformed state, the third image from the left in the bottom line shows that the speckle pattern changes and tends to disappear in the region of the strain localisation. This change in the structure of the pattern sometimes may lead to a break-down of the automatic evaluation at very high strains.

The break-down can be prevented if two subsequent sets of pictures are correlated to each other instead of correlating to the initial frame. This leads to an incremental evaluation of the strains. The draw-back of this incremental procedure is an increasing error.

The amount of pictures clarifies the high effort of a digital image correlation procedure. A subsequent specifically implemented scripting provides the stresses from the measured force values and the current width and thickness of the samples which are evaluated in the localisation zone with the smallest cross-section. The final results in terms of stress-strain-curves are then realised by the relation of the strain values to the forces and computed stresses, respectively.

6.3   Uniaxial Tensile Tests

Uniaxial tensile testing is a widely established and commonly standard test in the field of material characterisation. Typically, a tensile specimen is mounted on the device and loaded in axial direction. The load is applied either in force control or in displacement control. The raw data, obtained from the tensile test, is a force-displacement curve for the tested material. Using DIC the force-displacement curve is complemented by a series of pictures of the deforming sample. Different samples of the PBT with fibre contents from 0 to 30 wt% are used in the present investigations. The experiments are performed either on large tensile bars of type DIN EN ISO 527-2 type 1A [2] produced by injection moulding or on small bars according to DIN EN ISO 527-2 type 5A [2] cut out of the thin sheets. The advantage of tensile tests is that, on the one hand, they can be performed easily and, on the other hand, they gain a lot of insight into the material behaviour.

Different types of experimental procedures are realised with the tensile bars. The methods are explained in the following sections. First uniaxial tensile tests with different strain rates will be discussed. This will show the influence of the viscosity of the composite. In a second series of experiments, monotonous loading conditions with constant velocity are investigated until the samples break. The results are compared to uniaxial tests with several loading-unloading cycles. These tests gain detailed insight in the plastic behaviour and in the evolution of damage. All the uniaxial tensile tests on the composite material are realised with a device *Instron ElectroPuls E* 10000 at room-temperature in tensile-compression mode. The motion of the device is controlled directly by the displacement of the linear axis while the reaction force is continuously recorded by a force sensor mounted to the drive to obtain the measurement data, which are necessary for the evaluation of force-displacement curves and for the characterisation of the material. Evaluating the experiments in terms of stress-strain curves utilises the digital image correlation explained in the previous Sect. 6.2. This allows to calculate the local strains directly from the specimen's surface and, in addition, the true stress can be calculated from the force measurement

Fig. 6.13 360°-3D digital image correlation with four camera setup: tensile bar type 1A PBT inclusive necking area [18]

because the current size of the cross-section can be determined from the local strain information.

The deformation evaluation by means of the 360°-3D-optical deformation measurement in Sect. 6.2 is exemplary shown for a uniaxial tensile bar in Fig. 6.13. The presented DIC-results show a deformed state of the specimen at the beginning of the necking which is typically observed for thermoplastics. The 360°-3D setup is necessary for the simultaneous image capturing of the front and the rear side of the tensile bar allowing for a measurement of the current thickness.

The tensile bars are designed in such a way that the axial loading during a uniaxial tensile test can induce homogeneous stress and strain states in the centre of the sample under ideal conditions. Thermoplastics and in particular the investigated PBT show necking phenomena, i.e. even if the geometry of the specimen allows for homogeneous strain states, a pronounced strain localisation is observed in the experiments. Due to the temperature sensitivity of the material the localisation zone is typically observed in the region where the sample was touched during mounting. This localisation effect requires the full-field optical strain measurement because the exact position of the necking zone is not known a priori. Furthermore, the current size of the cross-sectional area changes in this region. While the engineering stress relates the measured force to the initial cross-sectional area the true stress is obtained as ratio

Fig. 6.14 Evaluation in the
localisation zone of the
uniaxial tensile bar type 1A
PBT

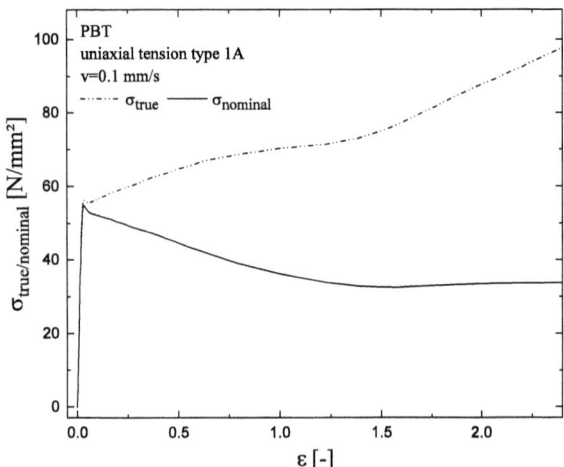

of the force to the current area. Figure 6.14 shows the engineering (or nominal) and
the true stress, which clearly differ if a strain localisation is observed. While the
engineering stress is constant after plastification or even decaying the true stress is
always increasing.

This effect is significant for the unfilled pure thermoplastic matrix material but
cannot be detected for the specimens with a sufficient high volume fraction of glass
fibres. The addition of short glass fibres changes the material behaviour of the composite from highly plastic to nearly elastic. The higher the fibre content in the composite material, the lower the influence of the remaining plasticity. The same can
be observed for viscous effects. While the unfilled PBT matrix shows a strong rate-
dependence the influence of the deformation rate is decreasing with increasing fibre
content. Both effects, the strain localisation and the viscous behaviour are not investigated further in this contribution because the are of low importance for the composites
with 20 and 30 wt% of fibres. However, these are the compositions of the composite
which are typical for industrial applications.

For the subsequent determination of the stress-strain curves, the maximum axial
strain $\varepsilon_1(t)$ at time t is extracted from the full-field strain measurement by the DIC
software. This maximum strain value is related to the local stress in the necking area.
According to Fig. 6.15, the index 1 concerns to the loading direction while indices 2
and 3 represent the width and thickness direction of the sample. The nominal stress
$\sigma = F(t)/A_0$ is obtained as ratio of the current force $F(t)$ at time t to the initial cross-
section A_0 while the true stress or Cauchy stress relates the same force to the current
cross-section $A(t)$, which is also determined from the DIC measurements. Due to the
production-induced anisotropy of the fibre orientation the current thickness and the
current width of the specimen have to be measured in the necking region. It cannot
be assumed that the transversal strains in width direction $\mathbf{e}_2$ and thickness direction
$\mathbf{e}_3$ are equal.

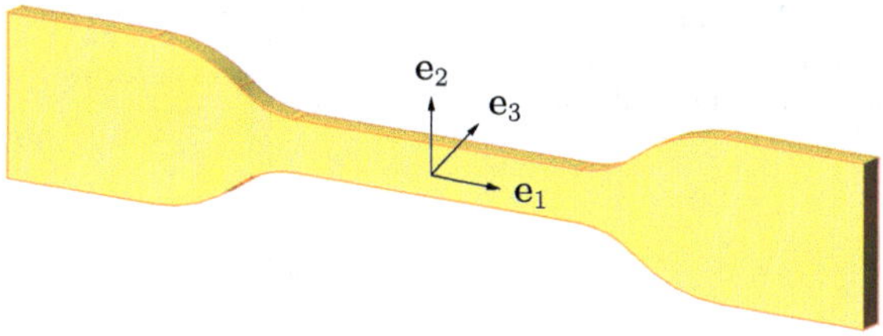

Fig. 6.15 Coordinate systems and axis directions of DIC (tensile bar type 5A)

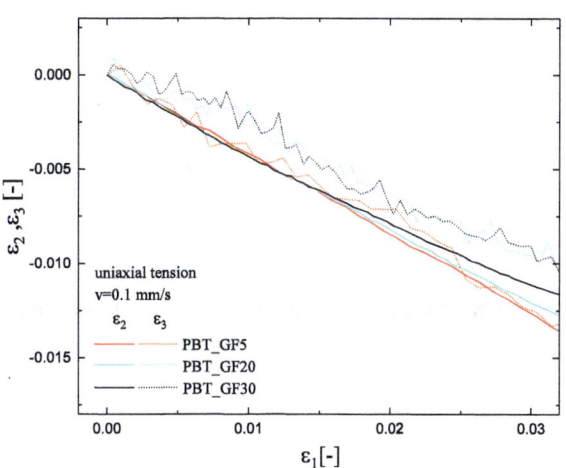

Fig. 6.16 Uniaxial tensile test (bar type 1A): strains orthogonal to the tensile direction

While the strains ε_1 in axial direction and ε_2 in width direction can directly be extracted from the DIC results, the thickness measurement in the 360°-3D camera setup requires the choice of so-called *gauge* points. This points are pairs of points, one on the front and one on the rear of the sample. The out-of-plane displacement at this *gauge* points is evaluated from the stereoscopic evaluation of the DIC results. Assuming homogeneity in the small thickness direction $\mathbf{e}_3$ of the sample allows to compute the strain ε_3 from the difference of the displacements at the *gauge* points and the initial thickness. Figure 6.16 shows the measured strains ε_2, ε_3 in the orthogonal directions as a function of the longitudinal strain ε_1. As can be seen, both strains slightly differ and the difference increases with increasing fibre content. The smooth curves in Fig. 6.16 represent the strain ε_2 while the oscillating curves show ε_3. The oscillations result from the local evaluation of the thickness change at the individual *gauge* points.

The evaluation of the digital image correlation allows not only for the analysis of anisotropy but also of compressibility. While elastomers are usually assumed to deform at constant volume and, therefore, behave incompressible, this is not true for thermoplastics. The three-dimensional strain measurement can also be used to determine the local volumetric strain. In a first step, the stretches λ_i are computed from the strains ε_i by

$$\lambda_i = 1 + \varepsilon_i \quad i = 1, 2, 3. \tag{6.5}$$

Fig. 6.17 Uniaxial tensile
test: volumetric deformation
as function of axial strain

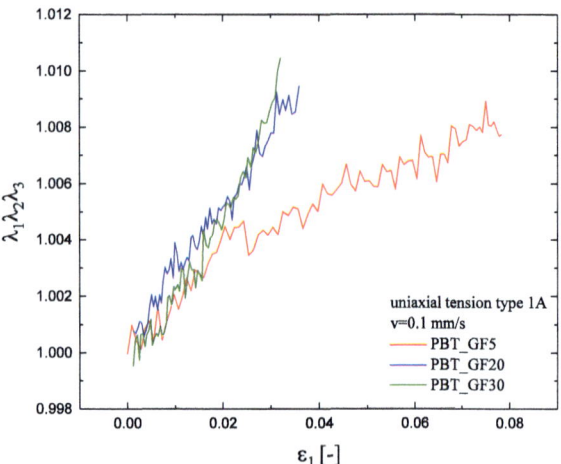

Figure 6.10 shows the deformation of a volume element. In a large strain setting the incompressibility constraint requires

$$\lambda_1 \lambda_2 \lambda_3 = 1. \tag{6.6}$$

Figure 6.17 shows the volumetric deformation as function of the axial strain for three different composites with fibre contents 5, 20 and 30 wt%. As can be clearly seen, the volume increases with increasing deformation and becomes more pronounced with increasing fibre content. Therefore, the composite cannot be assumed to behave incompressible. If an effective Poisson's ratio is computed as ratio of the transversal strains to the axial strain

$$\nu = -\frac{\varepsilon_2}{\varepsilon_1} \tag{6.7}$$

the observed value differs from $\nu = 0.5$ which represents incompressible behaviour. Note in passing, that in a geometrically linear framework the Poisson's ratio is a constant.

Due to the large deformations applied during the experiments, Fig. 6.18 shows a pronounce dependence of the effective Poisson's ratio on the axial deformation, which requires a large strain model in the simulations.

6.3.1 Uniaxial Tensile Tests at Different Strain Rates

For the chosen polymeric matrix material PBT rate-dependent effects are expected. Hence, uniaxial tensile tests at different strain rates with the pure matrix material have to be taken into account. Strain rates according to a cross-head speed of the

Fig. 6.18 Uniaxial tensile test: effective Poisson's ratio ν

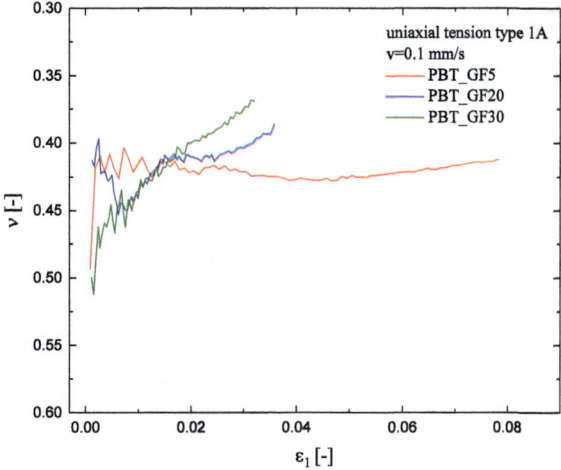

Fig. 6.19 Uniaxial tensile test: PBT at different strain rates

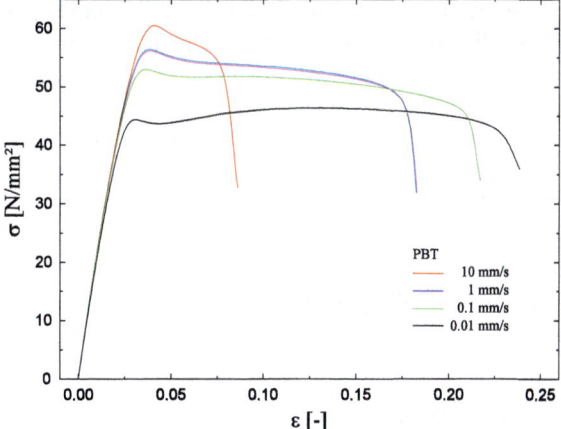

device of $v = 0.01$ mm/s, $v = 0.1$ mm/s, $v = 1$ mm/s and $v = 10$ mm/s are chosen. The specimens (tensile bar type 1A) are loaded until a necking area and the related softening becomes visible. The results of the PBT matrix are presented in Fig. 6.19. They show a pronounced visco-plastic behaviour of the semi-crystalline thermoplastic polymer. A yield point is clearly visible. The yielding takes place nearly without hardening. The formation of the localisation zone is related to the decrease of the nominal stress-strain curve after a first local maximum followed by small hardening. With increasing strain rate, the yielding point is shifted to higher stresses, the necking is attained earlier and the ductility of the specimens decreases, i.e. failure of the specimens occurs earlier at lower strains.

In Fig. 6.20 the stress-strain curves of the composite with a high fibre content of 30 wt% GF are shown for different loading velocities. The behaviour of the composite completely differs from the matrix material. On the one hand, the stresses

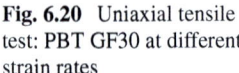

Fig. 6.20 Uniaxial tensile test: PBT GF30 at different strain rates

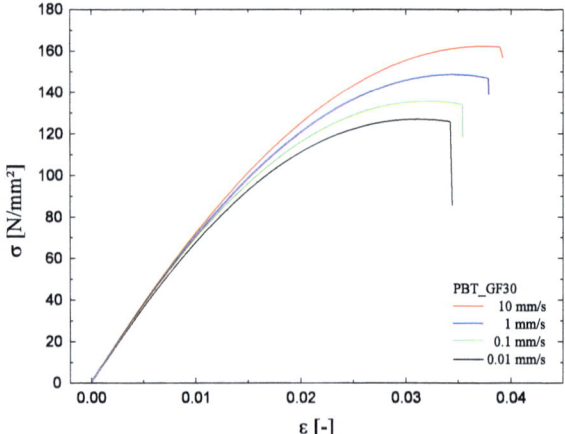

at comparable strains are much higher than for the pure matrix material, on the other hand, the ductility is reduced i.e. fracture occurs already at small strains. Even if a small rate dependence can still be seen, it is less pronounced than for the matrix material. In addition, a yield point and a large plastic region are not present, however, the failure of the samples occurs without showing strong softening effects.

In general, rate-dependent effects decrease with increasing fibre content. In the applications, typically higher fibre contents are in use. Therefore, the following comprehensive investigations are carried out at one constant strain rate related to the cross-head speed of the device of $v = 0.1$ mm/s. A detailed discussion of the rate-dependent effects is skiped.

6.3.2   Cyclic Loadings

Tensile Bar Type 1A

The analysis of loading-unloading cycles in the uniaxial tests of the composite material provides more information than a single loading path. Therefore, the results of the cyclic loading-unloading tests and the monotonous loading tests until failure are compared. The tests are performed using the moulded tensile bars of type 1A. The obtained stress-strain curves are shown in Fig. 6.21 for all available materials, i.e. the pure matrix and composites with 5, 20 and 30 wt% fibre content. Figure 6.21 summarises the results of the previous section. Increasing the fibre content increases the stiffness and strength on the one hand, but ductility and the strain at failure are decreased on the other hand. Furthermore, it is verified, that monotonous and cyclic tests lead to comparable stress-strain curves. Furthermore, the accumulation of damage can be seen in all unloading cycles as a decrease of the stiffness during unloading and reloading.

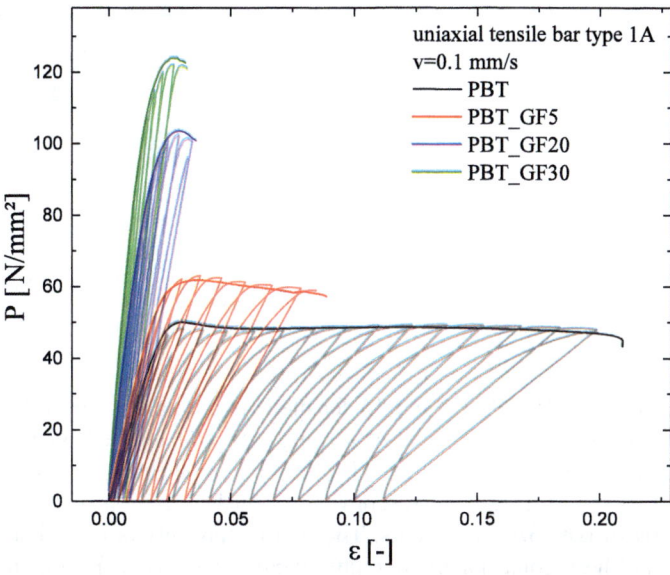

Fig. 6.21 Uniaxial loading until crack and cyclic loading tests (type 1A)

While the stress-strain curve of the monotonous test suggests nearly perfect plasticity of the matrix material, the cyclic test shows that possible hardening effects are compensated by the softening due to damage which occurs directly at the very beginning of yielding. Additionally to the matrix damage, the short fibre-reinforced composites allow for different types of damage mechanisms under load. In total the composite may undergo damage by fibre cracking, matrix cracking or debonding of fibres, which is sketched in Fig. 6.22. Macroscopically, in the analysis of the uniaxial tensile tests the difference between these damage mechanisms can not be determined. All these effects lead to a degradation of the stiffness during the unloading-reloading cycles as shown in Fig. 6.21. Material damage increases with increasing fibre content leading to the global decrease of the composite's ductility. Finally, the influence of the viscosity, which is neglected in further investigations as mentioned in Sect. 6.3.1, is reflected in the small hysteresis loops which is formed between unloading and reloading paths. The area of the hysteresis loops decreases with increasing fibre contents, which again confirms the assumption that viscous effects are negligible for the composite in a first approach.

Sheet Material

The fibres in the directly moulded tensile bars are mainly oriented in the axial direction. A variation of the fibre orientation is more or less impossible due to the production process but necessary to study the influence of the anisotropy on the mechanical behaviour. The influence of this process-induced main fibre orientation has been

Fig. 6.22 Types of damage in short fibre-reinforced polymers

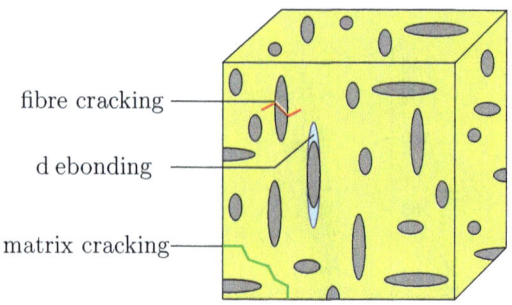

fibre cracking

d ebonding

matrix cracking

proven e.g. in [15]. Therefore, type 5A tensile bars are milled out of the available sheets with the thickness $d = 2$ mm. According to Fig. 6.6 tensile bars with different main fibre orientations can be obtained by variations of the milling angle β.

Figure 6.23 shows a light microscopy of a polished section of the rectangular cross-section of type 5A tensile bars. The first picture results from a specimen at $\beta = 0°$ while the second materialographic picture was taken from a specimen at $\beta = 90°$. While in the first case the fibres are mainly aligned in the tensile direction it can be seen that in the second case the fibres are aligned orthogonal to the tensile direction. These different main fibre orientations will lead to pronounced anisotropies especially for the specimens with high fibre concentrations.

In the following investigations four different angles β ($\beta = 0°$, $\beta = 45°$, $\beta = 60°$ and $\beta = 90°$) are chosen, which allow to study the influence of the fibre orientation explicitly. Figure 6.24 shows the DIC evaluation of the strain field using a two-sided camera setup before failure.

Figure 6.25 presents the results of the tensile tests on bars of type 5A in terms of stress-strain diagrams. Comparing stress-strain curves for different fibre orientations shows a distinguished anisotropic behaviour for the higher volume fractions of glass fibres. While the loading curves (a) and (b) for the pure PBT matrix and the low volume fraction of 5 wt% glass fibres show only a weak anisotropy the influence of the orientation increases with increasing degree of filling. This leads to larger differences between the stress-strain curves (c) and (d) for the different orientations. The weak anisotropic behaviour of the pure matrix material according to (a) results from the production process which induces an orientation of the polymer chains during moulding but this effect is negligible compared to the influence of the fibre orientation. In general the specimens with a main orientation of $\beta = 0°$ show the stiffest behaviour while the weakest results are found for the orientation of $\beta = 90°$. Only the loading path is investigated, the influence of the necking, which is mentioned above for the directly moulded tensile bars, is not studied further.

In Fig. 6.26 the loading and unloading curves are compared for the specimens of same orientation but different fibre contents. In (a) the orientation is chosen as $\beta = 0°$ which implies that the fibres are mainly aligned to the axis of the tensile bar. In this case the highest stress values are obtained for a given strain level. In contrast

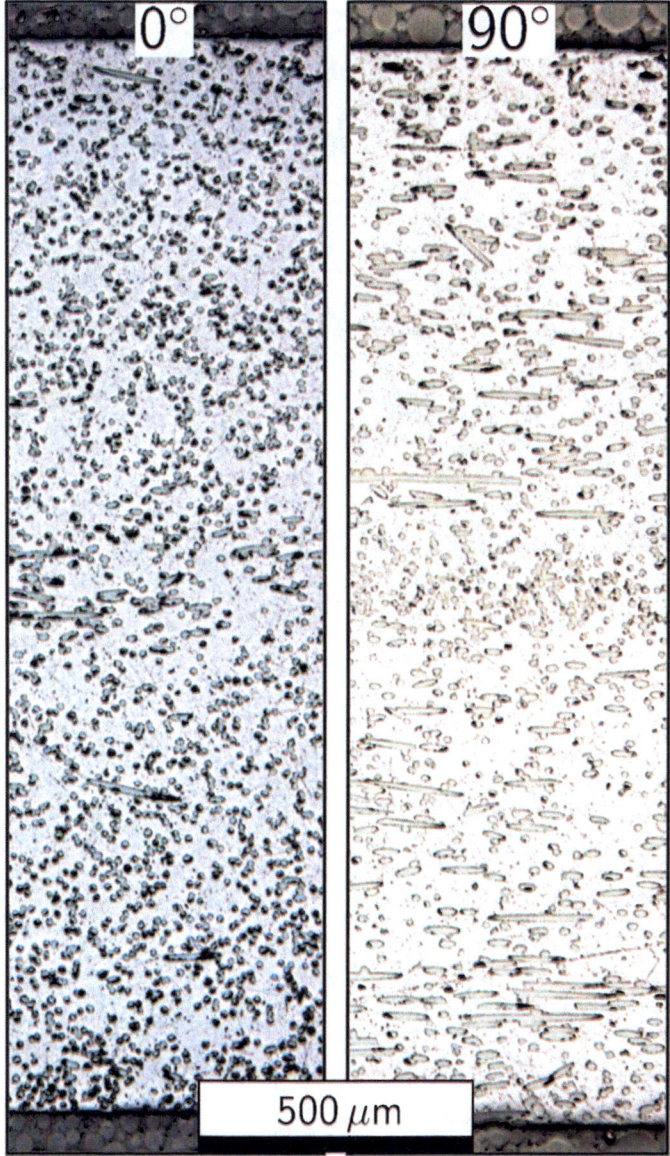

Fig. 6.23 Materialographic specimen: tensile bar type 5A PBT GF30

the results for a main orientation orthogonal to the axis of the tensile bar show lower stresses because the fibres are mainly aligned perpendicular to the loading direction. Both Figs. 6.25 and 6.26 show that the stiffness increases with increasing fibre content and with increasing alignment of the fibres with the loading direction. Furthermore,

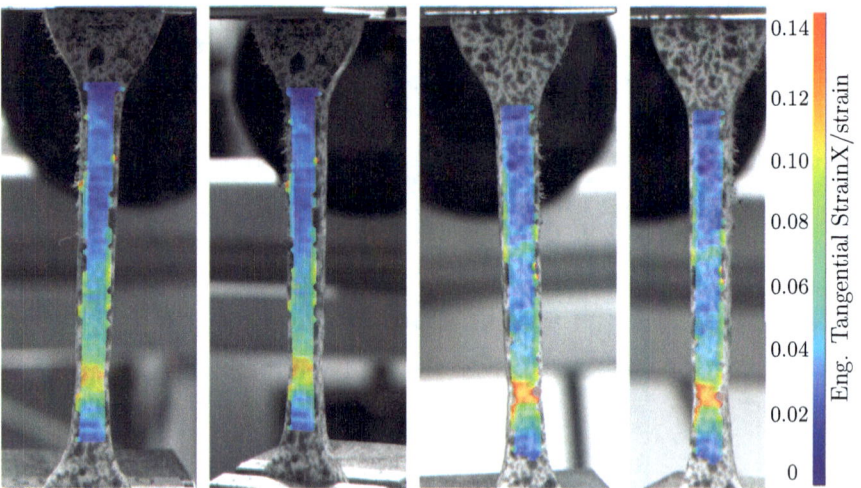

Fig. 6.24 DIC evaluation tensile bar type 5a

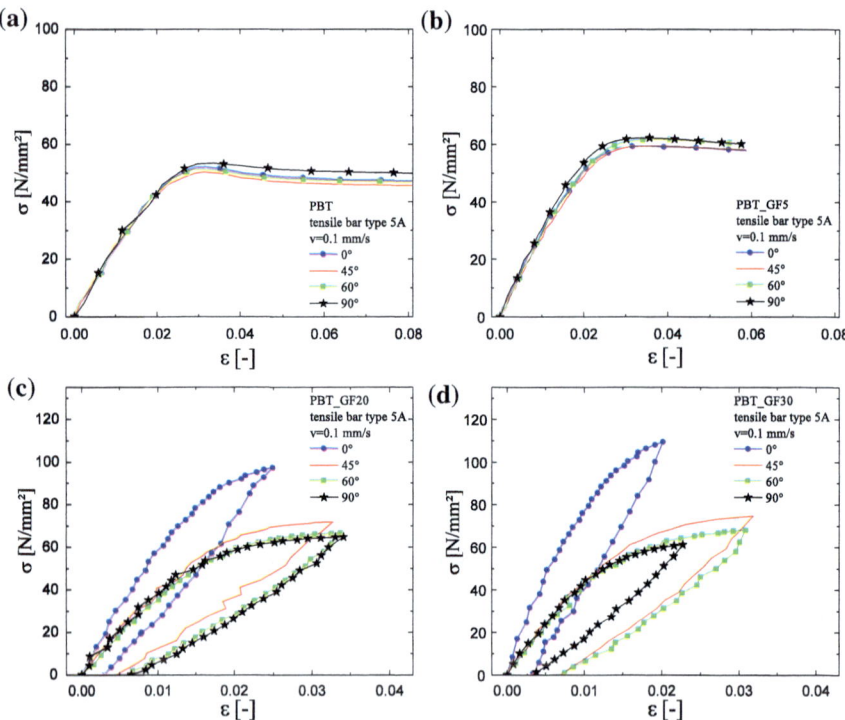

Fig. 6.25 Different glass fibre contents: **a** PBT, **b** PBT GF5, **c** PBT GF20, **d** PBT GF30

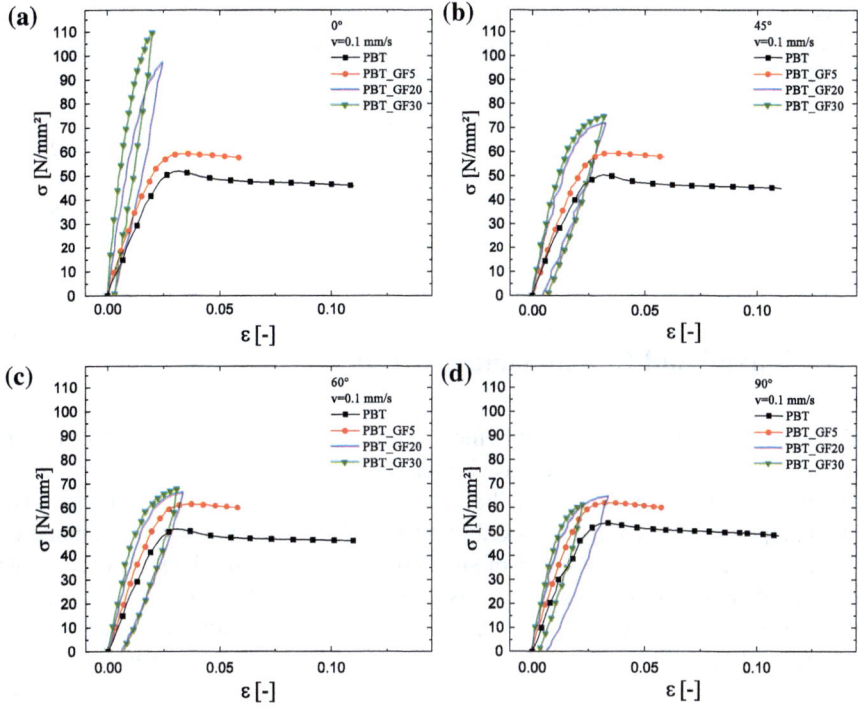

Fig. 6.26 Different glass fibre orientations: **a** $\beta = 0°$, **b** $\beta = 45°$, **c** $\beta = 60°$, **d** $\beta = 90°$

it can be seen, that the ductility of the matrix material is strongly reduced by the fibres. An increasing fibre content leads to decreasing plastic deformations and an early failure of the specimens at smaller strain levels for all orientations.

The results of the uniaxial tensile tests performed with specimens of different orientations prove the assumption of a strong directional dependency of the characteristic properties of the short fibre-reinforced composite material behaviour. The anisotropic behaviour influences the stiffness and the ductility of the samples. The effects increase with increasing fibre content.

6.3.3 Results of Uniaxial Tests

Based on the experimental results of the uniaxial tensile tests evaluated with the help of a digital image correlation system, the following material parameters are established. For the matrix material and the different fibre contents, the characteristic values summarised in Table 6.1 are extracted from the experimental analysis of the uniaxial tensile tests. The densities are taken from [25] and the Young's modulus is calculated as the secant of the increasing stress from 0 to 1 % of the nominal strain.

Table 6.1 Composite material parameters

Material	Fibre content w_f (wt%)	Fibre content n_f (vol%)	Young's modulus $E_{0.01}$ (MPa)	Density [25] ρ $\left(\frac{g}{cm^3}\right)$	Poisson's ratio ν
PBT	0	0	2330	1.3	0.4
PBT GF5	5	2.6	3130	–	0.42
PBT GF20	20	11.3	5960	–	0.43
PBT GF30	30	17.9	8240	–	0.45
Glass fibres	100	100	72,400	2.55	–

6.4 Biaxial and Inhomogeneous Tests

The uniaxial tensile test is probably the most widely used experiment for characterising the material behaviour because it is simple in its installation and performance. There exist a lot of standards for the uniaxial test. In contrast, it reflects the loading situation of real components only partially. The development of new material descriptions and the optimisation of simulation procedures need an even lager data set than obtained from the uniaxial tests. Previous works [23] have generally shown that the use of data from uniaxial tensile tests is not suitable for the calibration of a general material model which is supposed to describe the material behaviour under realistic multiaxial loadings. In an application the material usually does not only undergo uniaxial loadings but is loaded multiaxially with varying ratio of the principle stresses. Therefore, multiaxial characterisation methods are required if the model has to be calibrated for predictive simulations.

The biaxial test gives a possibility to obtain multiaxial data to evaluate the material modelling description. The setup of a biaxial test and its evaluation are much more complex than the uniaxial testing device. For the biaxial investigations in this work, a self-constructed testing device is used [21]. In contrast to the uniaxial setup there are two perpendicular axes that cross each other. Each axis consists of two drives which are moving in opposite directions. This guarantees that the cruciform specimen stays in an fixed position in the centre of the device. It furthermore simplifies the optical strain measurement by DIC and a stationary camera setup can be used. Both axes are controlled independently. Therefore, the specimen can be loaded independently in both directions and different stress states can be applied ranging from uniaxial tensile loading to equibiaxial tensile loading. As shown in Fig. 6.27 the forces are measured by force sensors mounted to the clampings on each axis. Note in passing, that the setup of the device does not allow for a direct evaluation in terms of stresses and strains due to its inhomogeneous nature but requires the solution of boundary value problems. A detailed description can be found in e.g. [11, 20].

For the biaxial testing device (see Fig. 6.28) a further specimen geometry, which is adapted to the existing setup and presented in Fig. 6.29, has to be extracted from the sheet material. Therefore, the samples are also milled out of the plates with a thickness $d = 2$ mm. The results of the uniaxial tensile tests presented in Sect. 6.3.3 show the

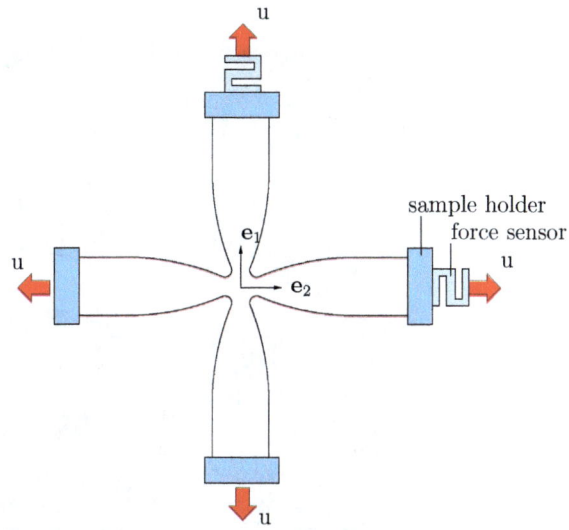

Fig. 6.27 Position of the biax specimen in experiment

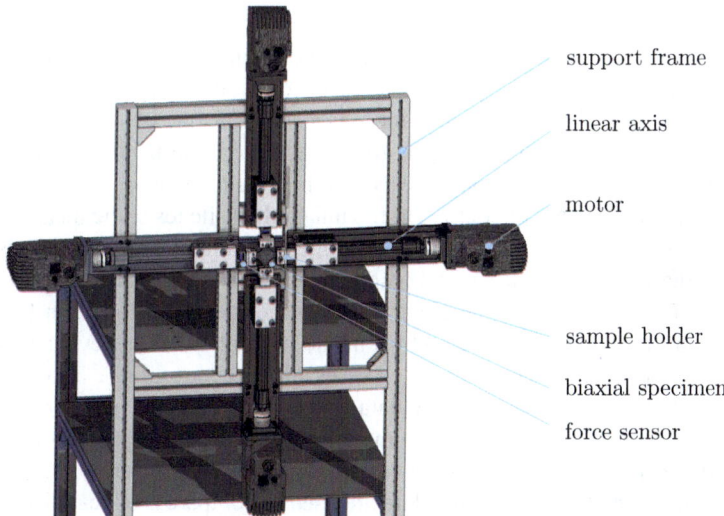

Fig. 6.28 Setup biaxial testing device [20]

strengthening of the composite with increasing fibre content. Therefore the maximum forces applied on the sample sleeves have to be high enough to cause a measurable deformation in centre of the cruciform specimen without inducing failure in the outer regions of the specimen close to the clamps. This is another limiting factor for the determination of a suitable specimen geometry. During the experiment the specimen

Fig. 6.29 Biaxial specimen

stays in a stationary position in the centre of the setup. Thus, the inhomogeneously deformed area in the middle is guaranteed for a continuous image recording during the loading steps. The evaluation of the images is performed by the DIC-software *ISTRA 4D®* giving the full field strain information on the surface of the specimen.

The interpretation of the biaxial tests and their results is more complex than for the uniaxial tensile tests. In contrast to the uniaxial tensile tests, the measured force values do not directly correlate to the specimen's strain and to the stresses in the centre of the biaxial sample. An inhomogeneous state is observed which has to be correlated to the motion of the clamps and to the measured forces in both axes. The procedure of an inverse method [7] has to be applied for the evaluation of the experiment in order to calculate the stresses in the centre of the biaxial sample. The field of strains is given from the data of the DIC-software. In order to perform the following explained procedure, a material model has to be chosen a priori. In Fig. 6.30 the inverse method is schematically presented to describe the procedure where experimental and simulated data are used to compare the local displacement fields. Starting the calculation, initially suggested parameters are needed, which can be taken e.g. from uniaxial tests. For the investigated glass fibre reinforced composite the fibre orientations have to be taken into account. The initially chosen parameters are subsequently optimised by minimising the difference between measured and computed strains on the surface of the whole sample. Finally, the stress field based on the measured strain values of the DIC can be calculated from the chosen material model with the optimised parameters [11, 20].

Equibiaxial tests with a speed of the clamping devices from $v = 0.1$ mm/s in each axis direction are chosen for the polymeric sheet material. The results of the biax-

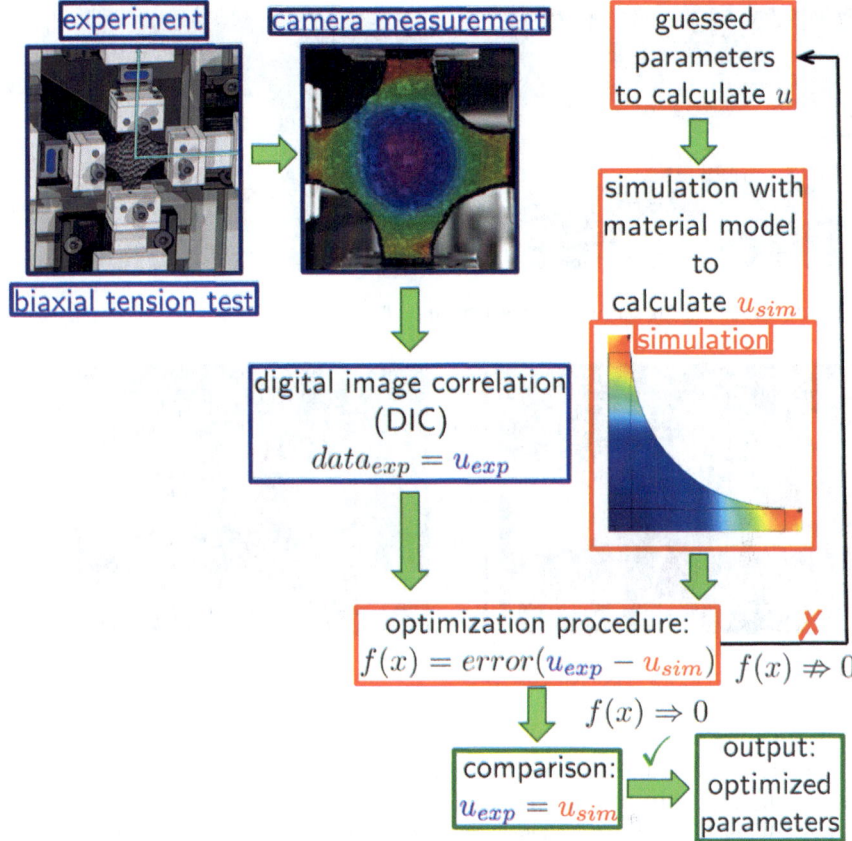

Fig. 6.30 Scheme of the inverse method

ial tests, presented in the following, are provided by the evaluation of the optical strain measurement system with the DIC-software. Therein the strains of the inhomogeneously deformed area in the middle are analysed. The values of the evaluated major strains (ε_1 and ε_2) in $\mathbf{e}_1$- and $\mathbf{e}_2$-direction are compared for the unfilled PBT in Fig. 6.31 and a fibre content of 20 wt% GF in Fig. 6.32.

For the unfilled matrix material, comparable forces and strains are found in both orthogonal axes directions. This again verifies the assumption that the pure PBT behaves nearly isotropic as was already motivated by the uniaxial tensile tests on the milled tensile bars of different directions (cf. Sect. 6.3.2). In the case of the PBT with a fibre content of 20 wt% GF the production process of injection moulding leads to an orientation, an alignment of the glass fibres parallel to the flow direction. This effect is also visible in the results of the biaxial experiments. Although the two axes are moving similarly, the axis direction of the cross specimen where the fibres are mostly oriented in parallel to the tensile direction is stiffer and deforms less even

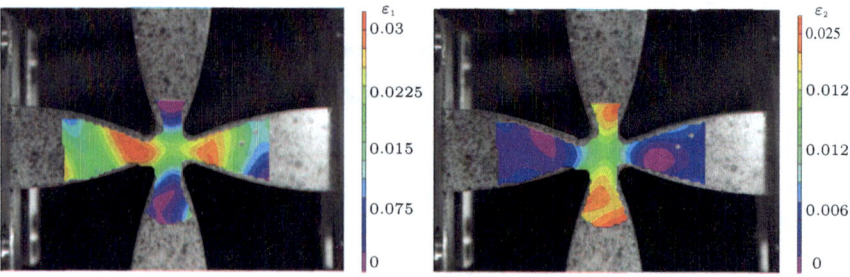

Fig. 6.31 DIC of a biax specimen of the unfilled PBT

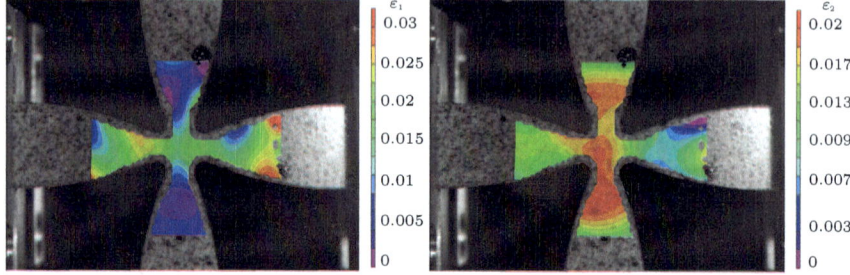

Fig. 6.32 DIC of a biax specimen PBT GF20

if the mean force values in this direction are higher. In contrast to that the other axis of the cruciform specimen shows lower force values. This effect is clarified in the DIC-evaluation results where the strains in the perpendicular directions are presented in Fig. 6.32. Here, the highest strain values are reached in the specimen axis and indicate the beginning of failure. Therein the axis parallel to a main fibre orientation is the stiffest one.

6.4.1 Nakajima Test

While biaxial devices as explained above are usually not available in a standard testing laboratory some alternatives for multiaxial examinations are established. One possibility is the so-called Nakajima test [16] which is approved for metallic materials where forming limit diagrams (FLD) have to be measured [5]. A uniaxial device is used to push a stamp into a thin plate which is clamped at its boundary. The test itself is comparable to a deep-drawing process and schematically presented in Fig. 6.33. Variations of the specimen's shape from full circles to small stripes allow to adjust the biaxial ratio in the centre of the specimen from equibiaxial to nearly uniaxial deformations. In the classical evaluation only the local strain in the centre of the specimen is taken into account to construct the FLD.

Fig. 6.33 Scheme Nakajima setup

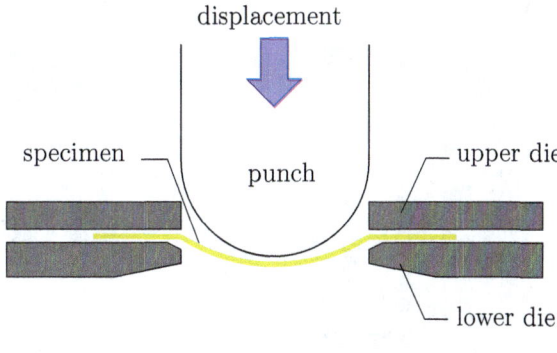

Fig. 6.34 Nakajima specimen $R = 60$ mm, $h = 150$ mm, $b = 150$ mm

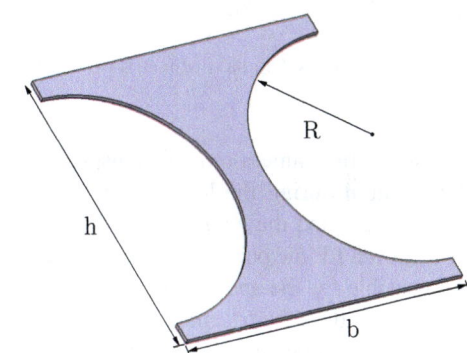

In this subsection, we describe the adaptation of the experimental setup for the fibre-reinforced PBT composite. The investigations of different loading situations with different biaxial ratios are realised by using different specimen geometries. First, circular specimens are milled from the sheet material (150×150 mm, $d = 2$ mm). Finally, these circles are tallied by two smaller circles. The shapes are based on those for metallic materials as proposed by Hasek [10]. For clarification Fig. 6.34 presents the specimen's geometry with a milled out radius of $R = 60$ mm.

With respect to the anisotropic behaviour there is the possibility to realise different loading situations by adjusting the fillet of the sample in relation to the process-induced main fibre orientation. In applicational case not only the perpendicular loading conditions are of of great interest for the investigated composite material. Therefore the influence of unknown multiaxial loading impacts in between is analysed. Additionally in order to give the possibility to compare the results of the Nakajima tests to the ones obtained from the biaxial tests different types of specimens are used. The complete circular plate ($R = 0$) is used and additionally different radii $R = 45$ mm, $R = 60$ mm and $R = 70$ mm are milled out from the sheet material.

Figure 6.35 shows the movable punch, which is mounted on a uniaxial device. It deforms the sheet material in the sample holder as shown in the detailed view in Fig. 6.36. An additional lubricant is applied on the surface of the punch in order to reduce the friction between the punch and the specimen. In the Nakajima setup a

(a)   (b)

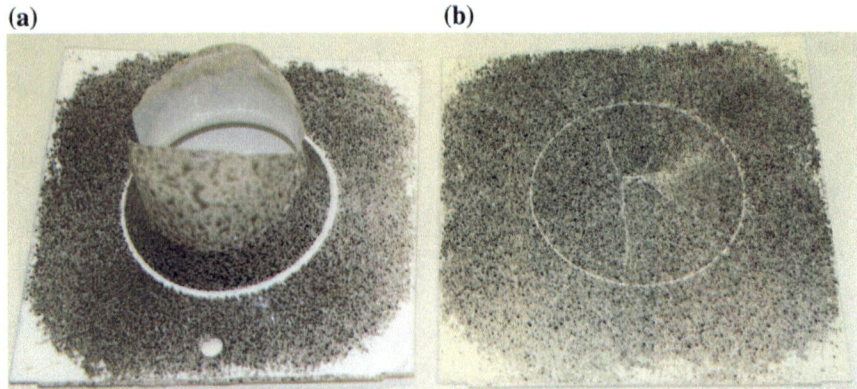

Fig. 6.35 Tested Nakajima specimen **a** PBT and **b** PBT GF30

one-sided four camera setup is indispensable to capture the photos of the specimen's deformation during the loading path. The maximum deformation depends on the fibre content and the geometry. While hemispherical states of the resulting cup can be achieved for the pure matrix material only small deformations with low curvature are possible for the specimens with 20 and 30 wt% fibre content. The deformation results including the fractures of the specimen are presented in Fig. 6.37 for the same experimental procedure. The ductile unfilled PBT shows the remaining hemispherical cup whereas the specimen with the high fibre content of 30 wt% on the right hand side deforms less ductile and breaks rather brittle. This is in line with the results found in the uniaxial test where the fibres strongly decrease the ductility.

In order to evaluate the experimental results of the Nakajima test the evaluation method of the biaxial tensile test has to be followed as explained in Sect. 6.4. In addition to the full-field strain measurement on the surface of the specimens the reaction force is also needed for the evaluation of the material modelling. Hence, a force sensor (max. 50 kN) which measures the values during the experiment continuously, is fixed at the punch while the test position u of the punch is controlled.

The examined multiaxial loading case that has been chosen is realised by the specimen geometry. The deformation of the complete specimen sheet provides an ideal equibiaxial tension while the uniaxial information is more and more approximated by a thinning bridge in the centred area (cf. Fig. 6.38).

The following results present the determined force-displacement curves, which are qualitatively correlated to a stress-strain curve. The velocity of the punch movement is $v = 0.1$ mm/s, like in the uniaxial tensile tests. Concerning the different fibre contents of the composite material, there is a significant characteristic material behaviour of each type of composite as shown in Fig. 6.39. The entire circular plate ($R = 0$) of the unfilled pure thermoplastic matrix material PBT shows the typical elasto-plastic behaviour with yielding, hardening and softening after localisation. Furthermore it shows the largest displacement values. These effects decrease with higher milled out radii. For each fibre content there is a decrease of the force as well

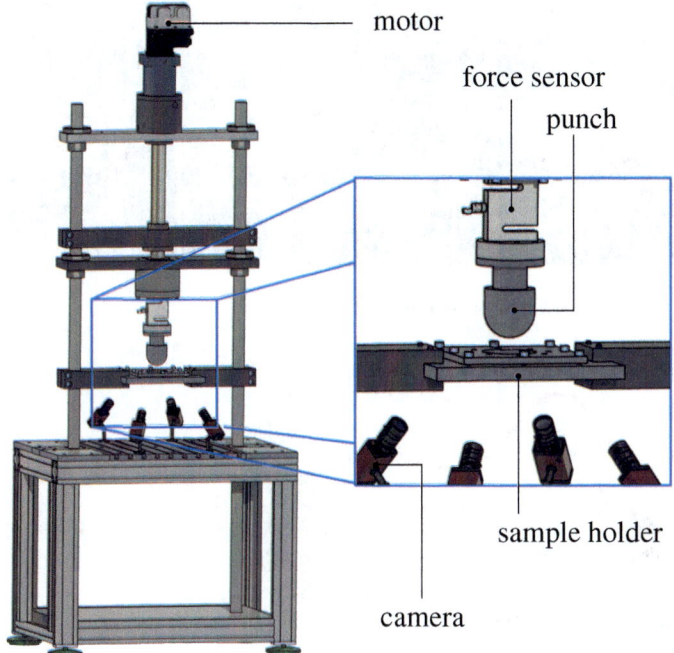

Fig. 6.36 Nakajima setup

Fig. 6.37 Detailed view:
sample in Nakajima setup

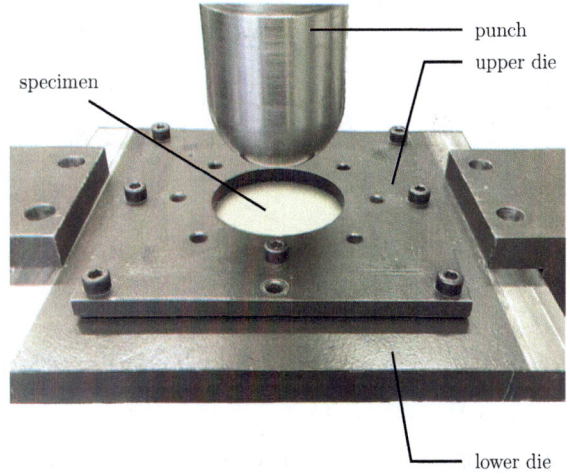

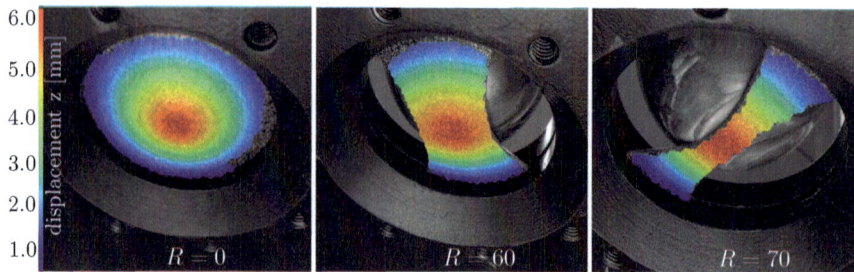

Fig. 6.38 Nakajima specimens with different radii

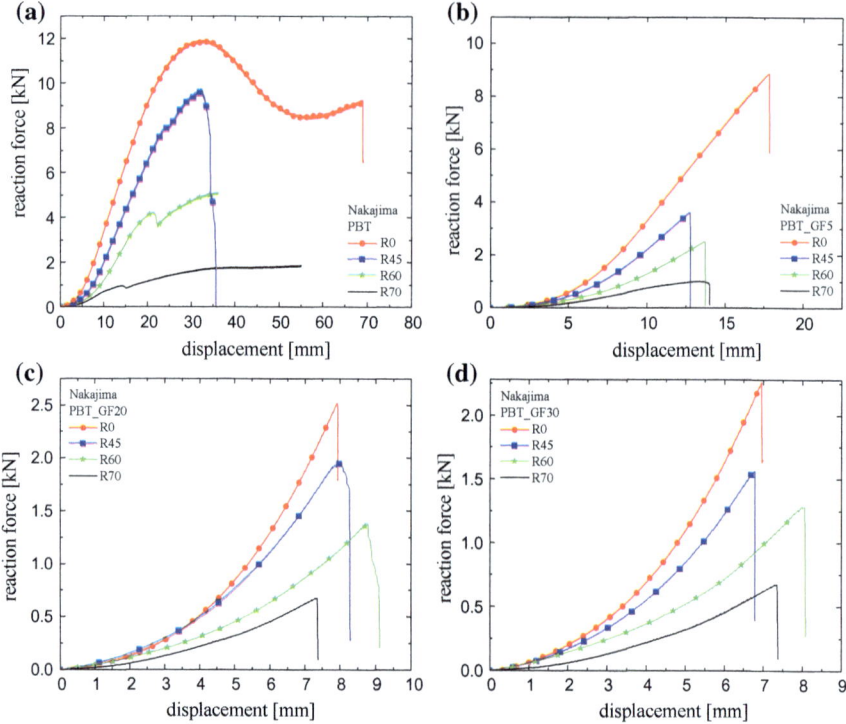

Fig. 6.39 Comparison of fibre contents: **a** PBT, **b** PBT GF5, **c** PBT GF20, **d** PBT GF30

as the displacement for thinner bridges of the specimen. Overall, the stiffness of the specimens increases with increasing fibre contents because the maximum forces are decreased significantly by a brittle failure.

The concluding results of the different specimen geometries are shown in Fig. 6.40. The influence of the remaining cross-section on the material behaviour is examined for each fibre content. For each type of specimen geometry there is the highest deformation for the unfilled PBT. The higher the fibre content the higher stiffness and

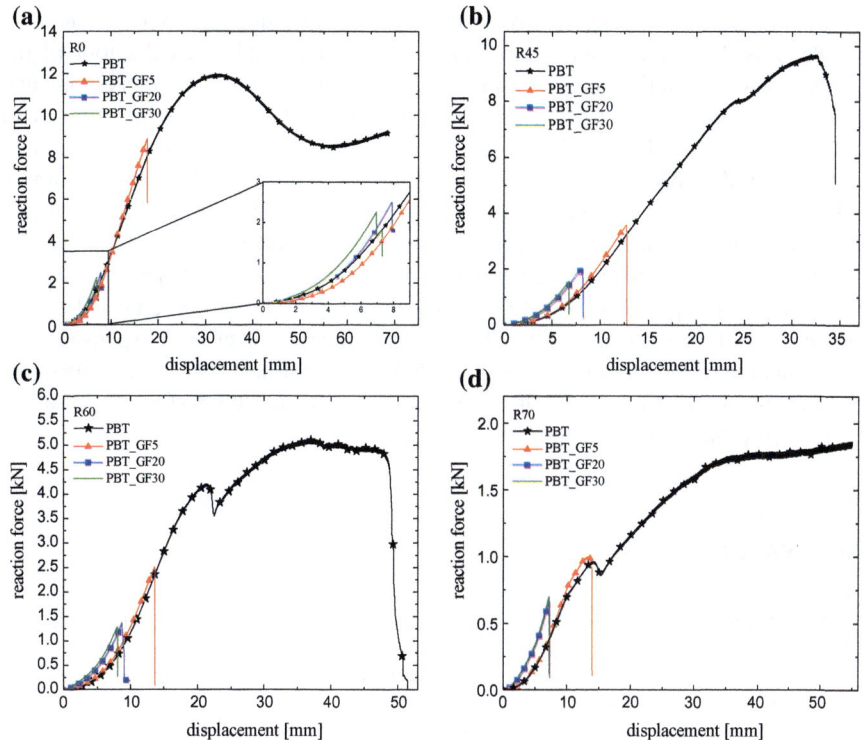

Fig. 6.40 Comparison of geometries: **a** PBT R0, **b** PBT R45, **c** PBT R60, **d** PBT R70

brittleness and the composite material fails at smaller displacements of the punch. Therein, the complete specimen without any radius reaches the highest forces related to an equibiaxial state of stress in the centre of the sample and the largest deformations. With an increasing radius size, i.e. with decreasing cross-section, the maximum forces of the loaded specimen decrease. There is nearly no difference in the results of the force values for the thin bridges with $R = 60$ and $R = 70$ mm in Fig. 6.40.

In addition the formability of a material can be investigated using the Nakajima test. From the correlated strain data concerning the principle strains are extracted by the DIC-software *ISTRA 4D®*. This information can be used to construct the classical forming limit diagrams (FLD) [1, 9, 24].

6.5 Conclusion

In this chapter experimental investigations have been presented in detail for the material characterisation of a short fibre-reinforced PBT matrix. The results give a comprehensive overview of the behaviour of the short fibre-reinforced PBT at room-temperature. The main effects have been carried out by different types of uniaxial tensile tests. It is found that the pronounced elasto-viscoplastic behaviour of the unfilled matrix material changes with increasing fibre content. Viscous effects and ductility decrease while stiffness and strength increase. Concerning the influence of a production-induced main fibre orientation further uniaxial and biaxial tensile tests followed by the Nakajima test have been realised. The results show the maximum stiffness for a loading in fibre direction. Furthermore, the anisotropy induced by the production process and already observed in the uniaxial tests also influences the results of the multiaxial tests. Each type of experimental analysis has been evaluated by means of an optical deformation measurement system with an attached digital image correlation software. The full-field strain measurement is indispensable for the evaluation of the inhomogeneous multiaxial tests. Further specific investigations concerning i. e. the thermal or chemical behaviour need to be performed for a full characterisation of the material.

References

1. Metallische Werkstoffe - Bleche und Bänder - Bestimmung der Grenzformänderungskurve - Teil 2: Bestimmung von Grenzformänderungskurven im Labor (ISO 12004-2) (2008)
2. DIN EN ISO 527-2, Kunststoffe - Bestimmung der Zugeigenschaften -. Teil2: Prüfbedingungen für Form- und Extrusionsmassen (1996)
3. Becker, T., Splitthof, K., Siebert, T., Kletting, P.: Error estimations of 3D digital image correlation measurements. Int. Soc. Opt. Photo. 63410ff (2006)
4. Botha, T.R., Els, P.S.: Digital image correlation techniques for measuring tyre-road interface parameters: Part 1—Side-slip angle measurement on rough terrain. J. Terramech. **61**, 87–100 (2015)
5. Campos, H.B., Butuc, M.C., Grácio, J.J., Rocha, J.E., Duarte, J.M.F.: Theorical and experimental determination of the forming limit diagram for the aisi 304 stainless steel. J. Mater. Process. Tech. **179**(1), 56–60 (2006)
6. Chu, T., Ranson, W., Sutton, M.: Applications of digital-image-correlation techniques to experimental mechanics. Exp. Mech. **25**(3), 232–244 (1985)
7. Cooreman, S., Lecompte, D., Sol, H., Vantomme, J., Debruyne, D.: Identification of mechanical material behavior through inverse modeling and DIC. P. Soc. Exp. Mech. Inc. **65**, 421–433 (2008)
8. Dantec Dynamics: Istra4D Manual (2015)
9. Geiger, M., Merklein, M.: Determination of forming limit diagrams—a new analysis method for characterization of materials' formability. CIRP Ann. Manuf. Technol. **52**(1), 213–216 (2003)
10. Hasek, V.V.: Investigation and theoretical description of factors relevant to the forming limit diagram-1. Blech Rohre Profile **25**(5), 213–219 (1978)
11. Johlitz, M., Diebels, S.: Characterisation of a polymer using biaxial tension tests. Part I: Hyperelasticity. Arch. Appl. Mech. **81**(10), 1333–1349 (2011)
12. Jung, A., Beex, L., Diebels, S., Bordas, S.: Open-cell aluminium foams with graded coatings as passively controllable energy absorbers. Mater. Des. **87**, 36–41 (2015)
13. Jung, A., Grammes, T., Diebels, S.: Micro-structural motivated phenomenological modelling of metal foams: experiments and modelling. Arch. Appl. Mech. **85**(8), 1147–1160 (2015)
14. Lucas, B.D., Kanade, T., et al.: An iterative image registration technique with an application to stereo vision. IJCAI **81**, 674–679 (1981)

15. Monte, M.D., Moosbrugger, E., Quaresimin, M.: Influence of temperature and thickness on the off-axis behaviour of short glass fibre reinforced polyamide 6.6 cyclic loading. Compos. Part A Appl. S. **41**(10), 1368–1379 (2010)
16. Nakazima, K., Kikuma, T., Hasuka, K.: Study on the formability of steel sheets. Yawate Tech. Rep. **264**, 8517–8530 (1968)
17. Pan, B., Qian, K., Xie, H., Asundi, A.: Two-dimensional digital image correlation for in-plane displacement and strain measurement: a review. Meas. Sci. Technol. **20**(6), 062,001 (2009)
18. Röhrig, C., Scheffer, T., Diebels, S.: Mechanical characterization of a short fiber-reinforced polymer at room temperature: experimental setups evaluated by an optical measurement system. Contin. Mech. Thermodyn. 1–19 (2017)
19. Scheffer, T.: Charakterisierung des nichtlinear-viskoelastischen Materialverhaltens gefüllter Elastomere. Dissertation, Universität des Saarlandes (2016)
20. Seibert, H., Scheffer, T., Diebels, S.: Biaxial testing of elastomers: experimental setup, measurement and experimental optimisation of specimen's shape. Tech. Mech. **34**(2), 72–89 (2014)
21. Speicher, K.: Konstruktion, Aufbau und Steuerung einer Biaxialanlage. Studienarbeit, Universität des Saarlandes (2009)
22. Sutton, M.A., Orteu, J.J., Schreier, H.: Image Correlation for Shape, Motion and Deformation Measurements: Basic Concepts, Theory and Applications. Springer Science & Business Media (2009)
23. Treloar, L.: The Physics of Rubber Elasticity. Oxford University Press, Oxford (2005)
24. Vacher, P., Haddad, A., Arrieux, R.: Determination of the forming limit diagrams using image analysis by the corelation method. CIRP Ann. Manuf. Technol. **48**(1), 227–230 (1999)
25. www.matweb.com: MatWeb—material property data. www.matweb.com

Index

© Springer-Verlag GmbH Germany, part of Springer Nature 2019
S. Diebels and S. Rjasanow (eds.), *Multi-scale Simulation of Composite Materials*,
Mathematical Engineering, https://doi.org/10.1007/978-3-662-57957-2

MIX
Papier aus verantwortungsvollen Quellen
Paper from responsible sources
FSC® C105338

Printed by Books on Demand, Germany